SUSTAINABLE CONSTRUCTION AND DESIGN

*The intention of this book is to nourish and sustain the Earth,
its inhabitants, and our future on this planet together.*

M. Regina Leffers, Ph.D.

*Director of the Center for the Built Environment
Associate Professor of Construction Engineering Technology
College of Engineering, Technology, and Computer
Science Indiana University Purdue
University at Fort Wayne*

Prentice Hall
Boston   Columbus   Indianapolis   New York   San Francisco
Upper Saddle River   Amsterdam   Cape Town   Dubai   London   Madrid
Milan   Munich   Paris   Montreal   Toronto   Delhi   Mexico City   Sao Paulo
Sydney   Hong Kong   Seoul   Singapore   Taipei   Tokyo

Editor in Chief: Vernon Anthony
Acquisitions Editor: David Ploskonka
Editorial Assistant: Nancy Kesterson
Director of Marketing: David Gesell
Senior Marketing Manager: Alicia Wozniak
Marketing Assistant: Les Roberts
Project Manager: Fran Russello
Art Director: Jayne Conte
Cover Designer: Bruce Kenselaar
Cover Photo: Scott Wargo/Oberlin College

Manager, Rights and Permissions: Zina Arabia
Manager, Visual Research: Beth Brenzel
Manager, Cover Visual Research
 & Permissions: Karen Sanatar
Image Permission Coordinator: Fran Toepfer
Lead Media Project Manager: Karen Bretz
Composition: Integra
Printer/Binder: Hamilton Printing Co.
Cover Printer: Demand Production Center
Text Font: 10/12 Palatino

Credits and acknowledgments borrowed from other sources and reproduced, with permission, in this textbook appear on appropriate page within text.

Library of Congress Cataloging-in-Publication Data

Leffers, M. Regina.
 Sustainable construction and design/M. Regina Leffers.
  p. cm.
 "The intention of this book is to nourish and sustain the Earth, its inhabitants, and our future on this planet together."
Includes index.
ISBN-13: 978-0-13-502728-8 (alk. paper)
ISBN-10: 0-13-502728-4 (alk. paper)
 1. Sustainable buildings—Design and construction. 2. Sustainable construction.
3. Leasership in Energy and Environmental Design Green Building Rating System.
I. Title.
TH880.L44 2010
690—dc22

                                        2009012972

10 9 8 7 6 5 4 3 2 1

Prentice Hall
is an imprint of

www.pearsonhighered.com

ISBN-13: 978-0-13-502728-8
ISBN-10:   0-13-502728-4

This book is dedicated to the Impulse to Flourish existent at the heart of every living thing in nature and to my brother, Daniel R. Leffers.

CONTENTS

Preface *x*
Introduction *xi*
Acknowledgments *xiv*

Part 1   The Foundations of Sustainability   1

Chapter 1   The Structure of Matter and the Material World   3
The Laws of Thermodynamics   5
The Ecology of Living Systems   6
Summary   7   •   Questions and/or Assignments   7   •   Notes   8

Chapter 2   Nature's Conscious Representatives   9
Summary   12   •   Questions and/or Assignments   13
Notes   13

Chapter 3   Generative Versus Degenerative Design   15
Summary   24   •   Questions and/or Assignments   24

Chapter 4   Whole Systems Thinking   25
Summary   29   •   Questions and/or Assignments   30
Notes   30

Chapter 5   Collaboration as Sustainability in Action   31
Summary   38   •   Questions and/or Assignments   39
Notes   39

Part 2   Sustainable Construction Roadmap   41

Chapter 6   Site and Natural Energy Mapping   43
USGBC LEED NC Certification Process   48
Case Example: Sweetwater Sound, Built to LEED NC Gold
Certification Level   48
USGBC LEED NC: Sustainable Sites (14 Possible Points)   50
Prerequisite 1: Construction Activity Pollution Prevention
(Required)   50
SS Credit 1: Site Selection (1 Point)   51
SS Credit 2: Development Density and Community
Connectivity (1 Point)   52
SS Credit 3: Brownfield Redevelopment (1 Point)   52
SS Credit 4.1: Alternative Transportation: Public
Transportation Access (1 Point)   52

SS Credit 4.2: Alternative Transportation: Bicycle Storage and Changing Rooms (1 Point)   53

SS Credit 4.3: Alternative Transportation: Low-Emitting and Fuel-Efficient Vehicles (1 Point)   53

SS Credit 4.4: Alternative Transportation: Parking Capacity (1 Point)   54

SS Credit 5.1: Site Development: Protect or Restore Habitat (1 Point)   54

SS Credit 5.2: Site Development: Maximize Open Space (1 Point)   55

SS Credit 6.1: Stormwater Design: Quantity Control (1 Point)   56

SS Credit 6.2: Stormwater Design: Quality Control (1 Point)   57

SS Credit 7.1: Heat Island Effect: Non-Roof (1 Point)   58

SS Credit 7.2: Heat Island Effect: Roof (1 Point)   58

SS Credit 8: Light Pollution Reduction (1 Point)   59

*Summary   60   •   Questions and/or Assignments   60
Notes   61*

Chapter 7   **Water Resources and Sustainable Landscaping   70**

USGBC LEED-NC Water Efficiency, Five Possible Points   71

WE Credit 1.1: Water Efficient Landscaping: Reduce by 50 Percent (1 Point)   71

WE Credit 1.2: Water Efficient Landscaping: No Potable Water Use or No Irrigation (1 Point in Addition to WE Credit 1.1)   71

WE Credit 2: Innovative Wastewater Technologies (1 Point)   72

WE Credit 3.1: Water Use Reduction: 20 Percent Reduction (1 Point); WE Credit 3.2: Water Use Reduction: 30 Percent Reduction (1 Point)   73

Summary   75   •   Questions and/or Assignments   75

Chapter 8   **Building Orientation, Renewable Energy and Storage, and HVAC Systems   76**

USGBC LEED-NC Energy and Atmosphere, 3 prerequisites and 17 possible points   78

EA Prerequisite 1: Fundamental Commissioning of the Building Energy Systems (Required)   78

EA Prerequisite 2: Minimum Energy Performance (Required)   79

EA Prerequisite 3: Fundamental Refrigerant Management
(Required)   79

EA Credit 1: Optimize Energy Performance 1–10 Points
(Two Points are Required for LEED-NC Projects
Registered after June 26, 2007)   80

EA Credit 2: On-Site Renewable Energy (1–3 Points)   82

EA Credit 3: Enhanced Commissioning (1 Point)   83

EA Credit 4: Enhanced Refrigerant Management
(1 Point)   83

EA Credit 5: Measurement & Verification (1 Point)   85

EA Credit 6: Green Power (1 Point)   85

*Summary   85   •   Questions and/or Assignments   86
Notes   86*

Chapter 9   Materials and Resources   87

USGBC LEED-NC Materials and Resources, 1 Prerequisite
and 13 Possible Points   88

MR Prerequisite 1: Storage and Collection of Recyclables
(Required)   88

MR Credit 1.1: Building Reuse—Maintain 75 Percent
of Existing Walls, Floors, and Roof (1 Point)   89

MR Credit 1.2: Building Reuse—Maintain 95 Percent of
Existing Walls, Floors, and Roof (1 Point in Addition
to MR Credit 1.1)   89

MR Credit 1.3: Building Reuse—Maintain 50 Percent of
Interior Nonstructural Elements (1 Point)   90

MR Credit 2.1: Construction Waste Management—Divert
50 Percent from Disposal (1 Point)   90

MR Credit 2.2: Construction Waste Management—Divert
75 Percent from Disposal (1 Point in Addition to MR
Credit 2.1)   90

MR Credit 3.1: Materials Reuse—5 Percent (1 Point)   90

MR Credit 3.2: Materials Reuse—10 Percent (1 Point in
Addition to MR Credit 3.1)   91

MR Credit 4.1: Recycled Content—10 Percent
(Postconsumer + 1/2 Preconsumer) (1 Point)   91

MR Credit 4.2: Recycled Content—20 Percent
(Postconsumer + 1/2 Preconsumer) (1 Point in
Addition to MR Credit 4.1)   92

MR Credit 5.1: Regional Materials—10 Percent Extracted,
Processed, and Manufactured Regionally (1 Point)   93

MR Credit 5.2: Regional Materials—20 Percent Extracted, Processed, and Manufactured Regionally (1 Point in Addition to MR Credit 5.1)   93

MR Credit 6: Rapidly Renewable Materials (1 Point)   94

MR Credit 7: Certified Wood (1 Point)   94

Summary   95   •   Questions and/or Assignments   95
Notes   96

Chapter 10   Indoor Quality—Air, Light, and Views   97

USGBC LEED-NC Indoor Environmental Quality, Two Prerequisites and 15 Possible Points   97

EQ Prerequisite 1: Minimum IAQ Performance (Required)   97

EQ Prerequisite 2: Environmental Tobacco Smoke (ETS) Control (Required)   97

EQ Credit 1: Outdoor Air Delivery Monitoring (1 Point)   98

EQ Credit 2: Increased Ventilation (1 Point)   99

EQ Credit 3.1: Construction IAQ Management Plan—During Construction (1 Point)   99

EQ Credit 3.2: Construction IAQ Management Plan—Before Occupancy (1 Point)   99

EQ Credit 4.1: Low-Emitting Materials—Adhesives and Sealants (1 Point)   100

EQ Credit 4.2: Low-Emitting Materials—Paints and Coatings (1 Point)   100

EQ Credit 4.3: Low-Emitting Materials—Carpet Systems (1 Point)   101

EQ Credit 4.4: Low-Emitting Materials—Composite Wood and Agrifiber Products (1 Point)   102

EQ Credit 5: Indoor Chemical and Pollutant Source Control (1 Point)   102

EQ Credit 6.1: Controllability of Systems—Lighting (1 Point)   103

EQ Credit 6.2: Controllability of Systems—Thermal Comfort (1 Point)   104

EQ Credit 7.1: Thermal Comfort—Design (1 Point)   104

EQ Credit 7.2: Thermal Comfort—Verification (1 Point)   104

EQ Credit 8.1: Daylight and Views—Daylight 75 Percent of Spaces (1 Point)   104

EQ Credit 8.2: Daylight and Views—Views for 90 Percent of Spaces (1 Point)   105

*Summary   106 • Questions and/or Assignments   107
Notes   107*

Chapter 11   Innovation and Design   112

USGBC LEED-NC Innovation and Design Process, Five Possible Points   112

ID Credit 1–1.4: Innovation in Design (1–4 Points)   112

ID Credit 1.1: Innovation in Design—Ice Storage/Reduce Peak Electrical Usage (1 Point)   112

ID Credit 1.2: Innovation in Design—Educational Building (1 Point)   113

ID Credit 1.3: Innovation in Design—Water Use Reduction 40 Percent (1 Point)   113

ID Credit 1.4: Innovation in Design—Heat Island Effect Roof (1 Point)   113

ID Credit 2: LEED Accredited Professional (1 Point)   115

Summary   115 • Questions and/or Assignments   116

Chapter 12   Sustainable Construction: A Collaborative Project   126

*Summary   134 • Questions and/or Assignments   134
Notes   135*

Chapter 13   Sustainable Construction Template   145

Index   147

PREFACE

The purpose of this book is to help the reader make the mental connections that are necessary to understand why building sustainably is an important thing to do. This understanding involves both the esoteric—the mental concepts of the world that we hold—and the practical—a different way of doing things in the practice of construction. Practical and esoteric information are not typically paired inside of the same book binding. Yet these two competencies of knowledge and action are completely interrelated, and a great disservice is committed by us when they are presented as if they are separate. Doing that creates an artificial presentation of the world, causing all manner of problems, one of which is exhibited in the way that we design and construct the built environment.

This book is for students in Construction degree programs, for those studying to take the United States Green Building Council's (USGBC) Leadership in Energy and Environmental Design (LEED®) Accredited Professional (AP) Exam, and for those who would like a better understanding of how to live a more sustainable life as conscious stewards of this Earth. Construction students are encountering more and more employers who enquire whether the students hold the LEED® AP designation before they are considered for employment. Within a few years, the USGBC's LEED® criteria have become the leading green building program in the United States, and is one that is being used all over the world as well. Its purpose is to identify many of the most important aspects of the construction process related to sustainable design and construction.

Through the experience of reading this material, my hope is that you will expand your ability to think about your own relationship with the Earth—what it is now, and what you would like it to become. Sustainable construction requires that we think about and conceptualize the world very differently than is our current practice. It also requires that those of us who are in the field of construction have a deeper knowledge of both the structure of matter (Physics) and the way that individual constructions of matter work together to create systems that support life (Ecology). This book explains the principles and methods of green building that emerge naturally when we employ biomimicry by consciously mimicking the processes of nature as we design and construct our habitats and buildings. You will read about examples of this way of building that already exist in the world, and will find clear and explicit how-to direction in sustainable construction.

Respectfully Submitted,
M. Regina Leffers, Ph.D.

INTRODUCTION

We are currently faced with global resource and ecological problems here on Earth. Actions that are taken "here," regardless of where "here" is, affect everywhere else too. It is intuitively obvious that air is not stationary—it is in constant motion, moving from one location to another all the time. The fouled air in one area gets inhaled by the inhabitants of its new location as it passes through. China's construction rate at present is so high that its people have to construct one coal plant per week in order to supply enough electricity to provide the new buildings with energy.[1] In April 2006, a satellite spotted "a dense cloud of pollutants" that moved from Northern China, crossed the Pacific Ocean, and reached the west coast of California, Oregon, and Washington. Researchers in these American coastal states found "specs of sulfur compounds, carbon and other byproducts of coal combustion coating the silvery surfaces of their mountaintop detectors."[2] These are the kinds of particles that when inhaled contribute to lung cancer. Buildings, commercial, industrial, or residential, no matter where they are constructed, consume more energy from fossil fuels and material resources than any other area of consumption;[3] hence, buildings are also the cause of a great deal of pollution.

The rainforests have been called the lungs of the planet, because they inhale such large amounts of carbon dioxide and exhale such vast quantities of oxygen. Another current example of a global problem is that those rainforests, our planetary lungs, are being cut down at an astounding rate. In the United States, we have created a way of living to which the people of many other nations aspire. But it would take more than the resources of our single planet Earth to provide enough energy to support and sustain the way in which we currently live *if everyone were to live as we do*.[4] And other countries are rapidly acquiring the ability to do just that. Type the words "carbon footprint calculator" (or "ecological footprint calculator") into a web browser and answer the questions on the quiz to find out how many planets it would take to provide enough resources to support and sustain your current lifestyle, if everyone on the planet lived as you live.[5] You can gain insight from the website as well about actions that you can take to reduce your own footprint. We must take individual actions now to reduce our footprint if we expect the world to be inhabitable for our children and grandchildren.

A conceptual paradigm shift is necessary at this time. If we hold the concept that the resources of the Earth are ours to do with as we please, and that we are wholly unconnected to them, then we use resources and discard them without any thought to the long-term consequences. Imagine the change that would take place if we held the concept that we are not only the stewards of the Earth's resources for present and future generations, but are also intimately connected to them, and that the choices we make about them can make the difference between destroying and benefiting the Earth for all of its present and future inhabitants.

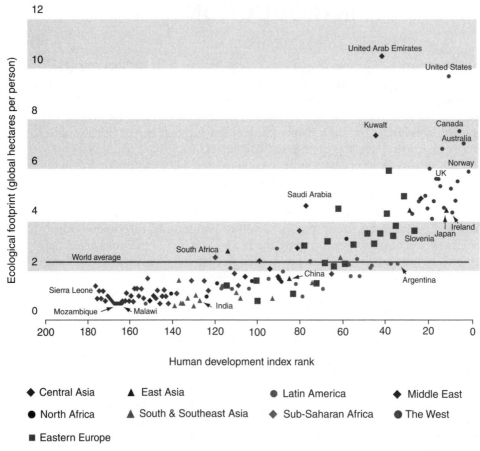

Human Welfare and Ecological Footprints Compared. *Global Footprint Network (2006);*
United Nations Development Programme (2006)[6]

Part I is about the origins of sustainable design, the ecological structure of matter, and the physical laws that govern it. Chapters 1 and 2 address the artificially separate way of conceptualizing the world that is prevalent, explores its origins in our culture, and presents some alternative ways of envisioning the world more holistically. These first chapters are heaviest on both the esoteric and physics side of the scale, but the material lays the foundation for the remainder of the book. So even though you're not used to encountering this kind of material in construction books, persist in working your way through them. In Chapters 3 and 4, we examine the difference in design if we approach that work from the perspective that everything is connected. What would the difference be? Biomimicry makes sense—but only if we place no artificial endpoint on our project design. We find that we must look at the future of our design work as well as the present. Chapter 5 takes a look at how we work together on projects if we keep in mind this actual fact of the material interconnection that exists, and proposes that collaboration is an adept outward expression of this reality.

Part II gives a sustainable construction roadmap. It views the construction process itself through this same lens of interconnectedness and gives some practical rules to follow to attain the goal of building sustainably in both commercial and residential applications. In order to gain an understanding of LEED® NC, we will follow a commercial project, scheduled to achieve LEED® NC Gold certification, Sweetwater Sound, through its LEED® credits and submittals. Chapter 6 looks at site and natural energy mapping and LEED® Sustainable Sites credits. Chapter 7 examines water resources and explains the LEED® Water Efficiency credits. In Chapter 8, we look at ways to orient structures on the site and methods of capturing natural energy, and we also look at renewable energy and storage methods and heating, ventilating, and air-conditioning (HVAC) systems. We also review the LEED® energy and atmosphere credits. Chapter 9 covers materials and resources and explains the LEED® MR credits. In Chapter 10, we look at indoor quality—air, light, and views—and explain LEED® indoor environmental quality credits. Chapter 11 uses the credits that Sweetwater Sound applied for in the category of innovation and design to help explain the LEED® innovation and design credits. In Chapter 12, we examine a current sustainable residential project, designed and built for Habitat for Humanity. Chapter 13 offers a helpful tool for thinking through the concepts involved in sustainable construction projects—a template that can be used until the process itself becomes second nature to you.

Notes

1. Peter Fairley, *Technology Review*, MIT, 2007.
2. Keith Bradsher and David Barboza, "The Energy Challenge: Pollution From Chinese Coal Casts a Global Shadow," *The New York Times,* June 11, 2006.
3. U.S. Department of Energy, Energy Information Administration, March 2001, *Monthly Energy Review*, and Lenssen and Roodman, 1995, "Worldwatch Paper 124: A Building Revolution: How Ecology and Health Concerns are Transforming Construction," Worldwatch Institute.
4. Rees, William E, (October 1992). "Ecological footprints and appropriated carrying capacity: what urban economics leaves out." *Environment and Urbanisation* 4(2): 121–130.
5. Wackernagel, M. (1994), *Ecological Footprint and Appropriated Carrying Capacity: A Tool for Planning Toward Sustainability.* Ph.D. Thesis, School of Community and Regional Planning. The University of British Columbia. Vancouver, Canada.
6. The Global Footprint Network (2006) and the United Nations Development Programme (2006) can be found at: http://Highlight_Findings_of_the_WA_SOE_2007_report_.gif.

ACKNOWLEDGMENTS

If it were not for the following people, I never would have been able to have the time and space to write this book.

Thanks to Dan Leffers, my brother and business partner, to Tim Hearld, my business partner and president of Synergid Commercial, to Jim Grim, Project Manager for Synergid, to Val Brelje, Office Manager and Job Cost Accountant at Synergid, and to Shelly Koelper, the CFO for both companies, Paul Davis Restoration of Northeast Indiana and Synergid, for their wholesome integrity, excellence, support, and friendship.

Thanks to Dr. Jihad Albayyari, Chair of the Department of Mechanical and Construction Engineering Technology at IPFW for his support and friendship.

Thanks to the students who put such incredible effort and enthusiasm into the Habitat for Humanity project: Adam Anspach, Anna Baer, Jennifer Bowman, Craig Campbell, David Carrion, Torrey Ehrman, Joseph Gabet, Rashid Hemed, Mary Kopke, Erica Lomont, Sam McClintock, and Adam Sordelet.

Thanks to the people who have read various versions of this manuscript and given me feedback, John and Beth Beams, Ann Beeching, Matt Leffers, Suzanne Katt, and Regina Sanders, and thanks especially to my husband, Mike Joker.

1

∎ ∎ ∎

The Foundations of Sustainability

1

■ ■ ■

The Structure of Matter and the Material World

Abstract

This chapter reviews the structure of matter and the first and second laws of thermodynamics—laws that govern the way matter behaves, regardless of size. Because the structure of matter and the laws of thermodynamics create the conditions that provide the foundation for living systems to grow and thrive, we must consider them when we build or construct anything. Sustainable construction and design requires that we understand these physical conditions that support biological flourishing so that we build with thoughtful intelligence.

This chapter begins with a quote from one of the most brilliant thinkers, mathematicians, and engineers of the last century, Buckminster Fuller. It is Fuller's exceptional mind that has given us one of the most lightweight, strong, cost-effective, and beautiful structures on Earth: the geodesic dome.

> "In short, physics has discovered
> that there are no solids,
> no continuous surfaces,
> no straight lines;
> only waves,
> no things
> only energy event complexes,
> only behaviors,
> only verbs,
> only relationships.[1]"

> *R. Buckminster Fuller*

Geodesic Dome located in Montréal, Québec, Canada.   Photographer, Terri Boake. Reproduced by permission.

This opening quote reveals one of the basic insights the science of physics has discovered about the way that matter itself is structured—that *all of life* at the level of subatomic particles *makes up a continuous whole.* When we look at the smallest particles of which matter consists, we can see that there is no actual separation in materiality. Subatomic particles are so completely interconnected, interrelated, and interdependent that they are not even able to be understood as isolated particles, but can only be understood within the context of their related phenomena as integral parts of a whole. Physics leads us to the understanding that everything that exists is a field of unbroken wholeness. This means that the classical idea of looking at and analyzing the world as if it were made up of individual separate parts that exist independently of one another isn't helpful. In fact, physics reverses the ideas about the world that we have held. We have always looked at the parts that appear to be individually functioning as the fundamental reality, and we've regarded the systems that these parts create as they work together as simply contingent arrangements of the parts. Instead, it is the "inseparable quantum interconnectedness of the whole universe" that is fundamentally real, and the parts that appear to be independent that are the contingent forms within the whole.[2]

The consequences of this interconnectedness on our efforts to conduct any scientific investigation of any particular event is that the particles constituting the material world can *only* be understood as an undivided whole, and that includes the particles of which "human beings, their laboratories, [and] observing instruments" are made.[3] Stated somewhat differently, thinking about any particle of matter as existing independently makes no sense at all. And we are made, as is the entire natural, material world, of these particles that must be viewed through the lens of quantum interconnectedness in order to be coherently understood—to make sense. That means that there is also no coherent concept of an independently existent construction design, site, project, materials, or resources in the built environment.

In our opening quote, Fuller tells us that physics has discovered that there are no solid things. Not only is there nothing that is actually solid, everything is an "energy event complex" made up of waves, behaviors, verbs, and relationships. What does that mean? It means that when we look at life in its smallest particles, we see molecules, atoms, electrons, protons, neutrons, neutrinos, and so on, moving around and interacting with each other. They do not stop at the periphery of the bark on a tree and begin again at the outer layer of the building standing next to it. They do not stop at the footers supporting the base of the building and begin again at the soil of the Earth's surface underneath. They are amassed together more tightly within the confines of the tree's bark or the building's skin, but they don't stop at the visible edges. These minute, active, and ever-changing particles that are the stuff of our physical, material earth continue along indefinitely, interacting and interrelating with all the other minute and active particles that make up the Earth and the universe. In fact, if our vision were ultramicroscopic rather than 20-20, it would be hard to discern where one thing ends and anything else begins.

THE LAWS OF THERMODYNAMICS

In this discussion of the structure of matter, we must also include the laws of physics that apply across the board, regardless of size or scale, to the way that matter works in our universe. Because of that, it is essential that we pay attention to these laws when we create or construct anything. Most of us have learned about them in school, but often, the way we learn about them seems unconnected to the practical world outside of school, and as a result, we don't keep them in mind when we design and construct buildings (or in the living of our everyday lives). However, whether we pay attention to them or not, they are not just theory—they are the actual, practical laws that explain how matter works in our world. They don't go away just because we ignore them. The laws of thermodynamics are akin to the laws of gravity; they apply to every particle and wave of matter that constitutes our universe, *from the perspective of life on Earth.*

The first law of thermodynamics states that energy cannot be created or destroyed. It can transfer from one state into another through processes like heating and cooling, but the amount of energy in the system doesn't change. The second law of thermodynamics is about entropy. It states that energy tends to disperse in large systems. The more solid and organized a material is, the lower the entropy value, and concomitantly, the more gaseous and disorganized the material, the higher the entropy value. What this means is that when we take something solid and organized like coal out of the Earth's crust and burn it, the carbon that would have taken eons to move from that solid substance and disperse into the atmosphere is changed into a substance that disperses in moments. That action creates an imbalance in our atmosphere that could change the ecologically supportive conditions that help human beings to be able to thrive on Earth.

THE ECOLOGY OF LIVING SYSTEMS

Understanding both the built and the natural environment through the lens of this existing interconnectedness (the structure of matter), and the laws of thermodynamics (the way units of matter work together), and its implied meaning that everything is an energy event complex, we can now see that every existing unit of matter in nature moves/grows/forms/changes, within this interconnected structure and following these laws. These are the embedded conditions of life on earth that contributes to nature's biological flourishing. Think about how the atoms, molecules, and cells of the human and animal bodies of the world exhibit these embedded conditions by interacting and interrelating with each other quite automatically—all of them working together toward the common goal of creating the living bodies that are you and me. Each minute, living part does its own individual job, to the best of its ability. We don't have to give directions or orders! No need to tell our hearts to beat or our lungs to go about the business of distributing the oxygen that we inhale. The entire work that goes on in the body is a creation of synchronous and intuitively interactive movement—a million individual elements, all striving toward flourishing!

We grew up thinking that our intelligence is located in our heads, and the brain has been regarded as the physical part of us that is in charge of the entire system. But biologists now tell us that conceiving of ourselves in this way is wrong-headed (pun intended)! It turns out that our cells are not just robotically living out a preprogrammed genetic code given to them by our DNA. In fact, the membranes of our cells have "receptor proteins" and "effector proteins" that respond to environmental signals with intelligence. In multicellular organisms like us, cells have responded to their environment to work efficiently and intelligently together. So for example, whereas a single cell "breathes" with its mitochondria, in our bodies billions of cells have specialized as a mitochondrial function and work together to form the lungs. And in the same way that a single cell moves through the interaction of two proteins called actin and myosin, the muscles in our bodies that enable us to move are created by communities of cells, which have enormous quantities of these proteins, working together collaboratively to fulfill the task of mobility.[4]

Think about the natural process of photosynthesis, and how the elementary units of nature all work together to create an organic movement that fosters prospering and flourishing of organic, biological units of nature. When trees use carbon dioxide from our planet's air and produce oxygen, it creates a process that allows us to live and thrive. But we must have plenty of trees in order for that to happen because we are biophilic—oxygen-breathing organic beings—and without the tiny cells in the leaves of the trees constantly turning sunlight and carbon dioxide into oxygen through the biological, organic process of photosynthesis, we would have nothing to inhale that would actually be able to sustain our lives.

Now think about how the planets, stars, and suns all move together in this same synchronous way. Think about the earth, and about all of the interactive processes that take place in order for life to continue along, grow, and develop here in this lush "garden of Eden" that we call home. Think about the simple action of our breathing and all that has to take place automatically, both inside of the biological unit that we think of as ourselves and outside of that biological unit, in order for our bodies to inhale and exhale. As we create the built environment, it is essential that we keep the interconnectedness of the material world and its working systems in mind in order to maintain an environment that is supportive of and sustaining to human life and to the land community with whom we co-inhabit this earth, our home.

Sustainable design and construction has at its core the specific purpose of finding ways of building that work within both the structure of matter and the laws that govern it.

Summary

In this chapter, we've looked at the structure of the material world through the lens of physics. The classical idea of investigating the material world as if individual parts exist independently from each other is revealed as being both misleading and wrong-headed. In fact, all of life at the level of subatomic particles makes up a continuous whole, and unless we view problems to be solved or knowledge to be gained through this understanding, we will achieve at best only partial solutions to problems, and knowledge that is only true under small given circumstances, but may have no relevance or truth at all when investigated in terms of the larger picture.

We also reviewed the laws of thermodynamics, which all of matter obeys, regardless of size. The first law of thermodynamics states that energy cannot be created or destroyed. It can transfer from one state into another through processes like heating and cooling, but the amount of energy in the system doesn't change. The second law of thermodynamics is about entropy. It states that energy tends to disperse in large systems. Solids have low entropy values, while gaseous material has high entropy values. We learned that we must think about these laws when we build or construct anything.

In the section The Ecology of Living Systems, we discussed the way in which matter is structured—the profound interconnectedness—together with the laws of physics that all matter obeys, as the embedded conditions of life from which biological flourishing is allowed to take place. Sustainable construction and design requires that we understand these physical conditions that support biological flourishing so that we build with thoughtful intelligence.

Questions and/or Assignments

1. What might be one important reason (in construction, or otherwise) to understand that the atoms and molecules of the world make up a continuous whole, rather than being disconnected?

2. Write a definition of the first law of thermodynamics.
3. Explain how the first law of thermodynamics relates to construction.
4. Write a definition of the second law of thermodynamics.
5. Explain how the second law of thermodynamics relates to construction.
6. What are some of the ecological conditions that are necessary for human life to continue on Earth?
7. Why is it important for someone who is working in the field of construction to understand the ecological conditions that support life?

Notes

1. Buckminster Fuller. *Intuition.* San Luis Obispo, California: Impact Publishers, [1970] 1983, Second Edition.
2. D. Bohm and B. Hiley. "On the Intuitive Understanding of Nonlocality as Implied by Quantum Theory," *Foundations of Physics*, vol. 5 (1975), pp. 96, 102.
3. David Bohm. *Wholeness and the Implicate Order.* London: Ark, 1983, p. 174.
4. Candace Pert. *Molecules of Emotion.* New York: Simon & Schuster, 1999.

2

■ ■ ■

Nature's Conscious Representatives

Abstract

Chapter 2 reviews the dominant role that the idea of being separate and unconnected from each other and from all of nature has been in our thought and culture historically. We discuss the endemic quality of this perspective, embedded as it is into our education in the form of "the scientific method." As the only biological systems in nature who have the ability to comprehend the consequences of decisions we make or actions we take, we bear a responsibility to future generations and to the environmental conditions that support all of life.

"In the end
We will conserve only what we love,
We love only what we understand,
We will understand only what we are taught."[1]

Baba Dioum, Senegalese Ecologist

The image of human beings unconnected to each other comes directly out of the culture that allows us to think about our mind and body as if they were separate and unconnected to each other. The dominant ideology in Western philosophy, thought, and culture has portrayed us as being isolated individuals and characterizes us as capable of being objective and detached observers. Neither of these conceptualizations are helpful and both have contributed to a condition in which we experience ourselves as being separate from and somehow the managers or owners of nature. Thomas Hobbes, a seventeenth-century English philosopher, expresses the idea of us as isolated individuals most clearly by recommending that we "consider men . . . as if but even now sprung out of the earth, and suddenly, like mushrooms, come to full maturity, without all kind of engagement to each other."[2] Hobbes got it wrong when it comes to mushrooms.

In reality, mushrooms are completely interconnected with each other. It's just very difficult to see that because the connection is only visible through a microscope. A mushroom is the above-ground "fruiting body" of an underground web-like network of microscopic mycelium, which connects the long root-like cells of (also microscopic) hyphae. In fact, the material that connects mushrooms to each other can grow for large periods of time without ever fruiting. There is actually a single fungus growing in Michigan that is a couple of hundred years old and is connected underground, extending throughout 40 acres.[3]

Hobbes got it wrong with mushrooms and he got it wrong with us human beings too. We don't spring out of the earth unconnected to others—we may spring out, but we do so out of a person—the interior of another human being. We are definitely not separate as we come into full maturity—we come into maturity as organic wholes in the context of our physical, emotional, mental, and peopled world with "all kinds of engagement to each other," as parts of various sizes of interconnected and interdependent organic wholes.[4]

In Western culture, we've used this idea that we are separate from whatever we are observing to acquire knowledge about our world—it's called "the scientific method," and every fifth-grade student learns about it in school. The idea that we are separate is embedded in our education in many ways. Moving from the perspective in which we think of ourselves as separate and objective—able to observe something from what we think of as a detached and disinterested standpoint—to the perspective in which we understand ourselves to be "energy event complexes" means that we also move into an internal position that makes splitting the body from the mind an impossibility—an inherent contradiction. We are one energy event complex within ourselves—one bodymind, and we are also one energy event complex within any knowledge bearing situation that we are investigating—always inside and never outside, always transactive and never simply interactive, always embedded and never objective. As an energy event complex, we are all of the thoughts that we think and *the field that encompasses who we are includes all that we think about.*

Transaction is a distillate word that means something far deeper and more complex than the word "interaction." Transaction means that we see everything both extensionally and durationally in its wholeness, *and that wholeness includes ourselves.* The word calls us to understand that what we look at is touched and changed by our observation, and that by our act of looking, we are also touched and changed by whatever we are observing. We human beings have evolved in nature as a part of nature, with intellect as a component of that physical evolution.[5]

Intellect has evolved in nature, as nature made conscious.[6] **We human beings are nature made conscious.** The terms *physical*, *psychophysical*, and *mental* are expressions that designate increasingly complex and intimate levels of transaction among natural events. The physical (matter) is made up of both living and nonliving events, and what marks the difference between them is that living things have needs, demands or efforts to satisfy those needs, and satisfactions. Living things experience an unstable or uneasy equilibrium. When they sense imbalance, they actively interact with their environment to regain their ordinary, internal condition of balance, and because of that, their customary equilibrium is restored.[7]

When organized transaction between feeling creatures is actualized in communication and language, we add the property of mind. With this property added, the organism is able to do more than just have feelings; the feelings have sense and they make sense. Language enables us to discriminate, identify, and communicate feelings. It is "the link between experience and existence,"[8] and "it is every bit as much a consequence of our natural history as the emergence of the human hand or brain; like them its forms arise as progressive adaptations of the organism to its environment."[9]

Mind has developed *in* the world—"a body-mind, whose structures have developed according to the structures of the world in which it exists."[10] In the word "bodymind," body refers to the part of the organism which is continuous with all of nature, whether inanimate or animate; and mind refers to the characteristics that emerge when the organism is involved in more complex and interdependent situations. The functions of mind have developed directly out of the organization and patterns of organic behavior.

What we think matters. Our thoughts are made up of atoms and molecules that are not only a part of us, but are interconnected with whatever we are thinking about.

What we feel matters. Our feelings are also made up of atoms and molecules that are both a part of us and transactive with whatever we are having feelings about.

What we do matters, and concomitantly, *what we don't do matters.* Our actions carry the force of all of the atoms and molecules that are involved in that action. Any action we take matters.

Philosophers through the ages have maintained that *we confer value onto that which we choose by the act of our choosing*—this is essentially why it is so important that we understand the gravity of the fact that we are nature's conscious representatives. We are the only biological beings in nature who have both the ability to choose and the ability to be conscious of the choices we make. Whether that choice is expressed and acted on through our thoughts, our feelings, or our actions, it is the composite of these choices that are the details of who we are as well as who we are becoming. The choices we make are the face of our soul lived in this world. They are the lived expression of our character, as we have, and as we are creating it. As the "energy event complexes" that we are, everything that we think, feel, or do touches all that is, with the force of its valuing presence, impacts that which it thinks, feels, or acts upon/toward, and is changed because of it.

Our conceptions of self and other provide the ground for us from which we make the valuing decisions that determine what we do. That is why a conceptual paradigm shift is necessary at this time. If we hold the concept that the resources of the earth are ours to do with as we please, then we may use and discard them without any thought. Imagine the change that would take place if we held the concept that we are the stewards of the earth's resources for present and future generations, and also for the benefit of the earth itself and for all other inhabitants, the "land community." When we create something with no thought about what will become of it when it reaches the end of its usefulness, "cradle to grave" is an apt name for that process. When we're done with it (whatever *it* is), we throw it away and it goes to its grave at the dump. There is a better way of creating things, using what can be described as cradle-to-cradle thinking.[11] We can plan into the design the next life of the thing when it reaches the end of its usefulness in its current incarnation. The terms *generative* and *degenerative* can also be used to more fully describe the differences between these two design approaches.

Historically, we have used the Earth's resources as if they were separate from our living, breathing, thinking selves. It is easy to feel that disconnection from nature at this juncture in our human story. It is important for us human beings to reconnect to nature on all levels of being and action, and to feel that reconnection. It is easy to feel overwhelmed by the implications that arise when we actually take in the knowledge that we are continuous with nature. However, being overwhelmed doesn't help anything. We human beings usually embed even more deeply in our habitual behaviors when we're overwhelmed, trying to find a way to comfort ourselves. When those very habits are the ones that cause the problems, it makes us realize just how large of a change is being called for, and that is what we're after here.

One way that we can combat the feeling of being overwhelmed is to begin on a small scale. It is necessary for us human beings to walk in nature—to go out to the land, to cultivate and care for it. There are many small ways in which we can fulfill this human need. We can take a part of the yard to create a garden and grow some of our own food. Or if we live in the city and have no yard available, then we can use a window box or container in which to garden. We could do something that has been done in other cities and help to start a neighborhood garden. We could volunteer in the park nearest our home, or we could do something on a larger scale and use summer vacation time to help care for the trails in our national park system. When we walk on the land, every part of our being feels the ground underfoot, the sun and the wind, and the air on our skin. The feeling of connection that those sensations give us is just plain difficult to feel in a city made of concrete, asphalt, and steel.

As the conscious representatives of nature, we are inseparably connected to and are part of this environment that we must steward. If we are harming the environment, we are harming ourselves. If we take care of the environment, we are taking care of ourselves as well. We must be as concerned about a polluted stream for the sake of the life and health of the stream as we would be concerned about a polluted artery in our own bodies. The connection between the health of the physical environment and our own health is intimate—we are an intimate part of the physical environment *and* the physical environment is an intimate part of ourselves.

Summary

In this chapter we've reviewed the dominant role that the idea of being separate and unconnected from each other and from all of nature has been in our thought and culture historically. This is the perspective or ideology from which springs the experience that human beings are both separate from and the owners or managers of nature—the underlying framework that seems to give us permission to use material resources as we please without any thought to how it may affect future generations or the health of the environment. This perspective is embedded in our educational process in "the scientific method," the mode of observing and acquiring knowledge about our world.

We review the implications of the physical condition of there being no actual separation of material particles when viewed in their smallest components on knowledge-bearing situations. As the only biological systems in nature having the ability to think and to consciously make decisions, we human beings must consider outcomes before we act. Historically, we have used the Earth's resources as if they were separate from our living, breathing, thinking selves. It is important for us to reconnect to nature on all levels of being and action and to feel that reconnection. The connection between the health of the physical environment and our own health is intimate—we are an intimate part of the physical environment and the physical environment is an intimate part of ourselves.

Questions and/or Assignments

1. Contrast the actions of someone who exhibits the mentality of "ownership" of the earth's resources (building materials) with that of someone who exhibits the mentality of "stewardship" of those same resources.
2. What is the design concept of "cradle to grave"? Explain how that concept relates to construction.
3. Describe a "cradle to grave" treatment of one construction material/resource. Write a history of its life from beginning to end, using a time-line format.
4. What is the design concept of "cradle to cradle"? Explain how that concept relates to construction.
5. Using the same material/resource from your answer to question 3, write an alternate history for it using the "cradle to cradle" ideology, again using a time-line format.
6. The author states that "the connection between the health of the physical environment and our own health is intimate." Give an example of this from your own experience.

Notes

1. Baba Dioum, Senegalese ecologist. He is the General Coordinator of the Conference of Ministers of West and Central Africa, and organization that represents 20 African countries. This quote is taken from a speech made in New Delhi, India, to the general assembly of the International Union for the Conservation of Nature.
2. Thomas Hobbes. "Philosophical Rudiments Concerning Government and Society," in *The English Works of Thomas Hobbes*, ed. Sir W. Molesworth, [1651] 1966, vol. II, p. 109.
3. Michael Pollan. *The Omnivore's Dilemma*. New York: The Penguin Press, 2006, p. 375.
4. Jane Addams. *Twenty Years at Hull-House*. New York: The Penguin Press, [1910] 1981, pp. 98–100. This concept of "people as organic wholes" is an underlying ideal in much of Addams's writing.

5. John Dewey and Arthur F. Bentley. *Knowing and the Known*. Boston: The Beacon Press, 1949.

6. John Dewey. "Experience and Nature," in *The Later Works of John Dewey* 1925–1953, vol. 1, ed. Jo Ann Boydston. Carbondale: Southern Illinois University Press, 1988.

7. Dewey. See especially Chapter 7, "Nature, Life, and Bodymind."

8. R. W. Sleeper. *The Necessity of Pragmatism: John Dewey's Conception of Philosophy*. New Haven: Yale University Press, 1986, p. 107.

9. Sleeper. *The Necessity of Pragmatism*. p. 205.

10. Dewey. *Experience and Nature*. p. 211.

11. William McDonough and Michael Braungart. *Cradle to Cradle*. New York: North Point Press, 2002.

3

■ ■ ■

Generative Versus Degenerative Design

Abstract

Chapter 3 reviews the difference between generative and degenerative design as it applies to the built environment. It describes a new design leadership that has emerged in construction-related fields that mimics the generative design that occurs in nature. The idea of designing and building structures that are able to use and replenish energy, light, and water in a way that mimics nature is not new, and is achievable. A home designed in the 1920s by Buckminster Fuller went a long way toward showing us how to do this. On Earth, all of nature is generative—every single iota in nature has a next purpose in the system. We can build our structures with this same feature, so that they too have a next purpose when their first life is complete.

> "The world, we are told, was made especially for man—
> a presumption not supported by all the facts.
> A numerous class of men are painfully astonished
> whenever they find anything, living or dead,
> in all God's universe, which they cannot eat or render
> in some way what they call useful to themselves."
>
> *John Muir*

It is because of John Muir and his love for nature in all of its wildness that we have some of our largest nationally owned acres of natural wild park area, which we get to enjoy. It is primarily because of forward-thinking people like him that we have a tradition of preservation in the form of our national, state, and local park systems. John Muir is the person who founded the Sierra Club, a preservation movement that even today continues to grow and thrive. He believed that one of our most sacred duties as human beings is to protect and preserve nature. I agree with John Muir, and as you already

John Muir, American Conservationist.
Taken by Professor Francis M. Fritz in 1907.

know from reading the previous chapters, my belief in that duty comes from the fact of our being the conscious representatives of nature.

As we drive from place to place across our country and witness our rolling hills, farmland, mountains, deserts, and plains being continuously flattened to make way for more housing additions and strip malls, we can appreciate the foresight of those who have led the movement to preserve some of our greatest natural beauty in the form of national, state, and city parks. In terms of the building industry, we now have another form of leadership emerging that will help to ensure that the quality of our air, water, and other natural resources in the built environment stays healthful and life-giving. One such example is happening in Portland, Oregon. There is a 35-block area of downtown Portland, Oregon, called Lloyds Crossing, which is being designed to mimic the strategies employed by the generative design that happens in nature's forest. The collaborators in the design process did something that is so simple that we are left asking ourselves why we don't do this every time we develop or build anything!

Typically, folks in the industry who are consciously striving to repair our environment choose a baseline that includes the human-built environment as it is currently, or as it was at some fairly recent date in history, and try to improve the existing conditions ecologically (usually meaning reducing carbon emissions) to that point in time. The designers of Lloyds Crossing did something much more radical. The baseline they used to start from was the undeveloped land as it was before human beings held the concept of land ownership and

Lloyd's Crossing Perspective Drawing.
*MITHUN, Seattle, WA. Reproduced by
Permission.*

started to establish permanent structures. They looked to the mature conifer forest ecosystem that had been growing there and measured, among other things, how much water it would have absorbed, how much sunlight and carbon dioxide it would take in, and how much oxygen it would have generated. These are the measurements they used as the baseline to work toward in their goal of achieving carbon neutrality by the year 2050. They will increase the development density to 10-million square feet, and will make the carbon footprint of the area that of the original conifer forest at the same time.

In valuing water as the precious resource that it is, the first design goal they worked toward was to create a neighborhood that *lives within its "water means."* The water that falls there will be used, treated, and reused there.

Lloyd's Crossing Plan View.   *MITHUN, Seattle, WA. Reproduced by Permission.*

They calculated that 64-million gallons of rain per year falls onto this 35 block area. Fifty percent of the rainwater that fell onto the area when it was a conifer forest soaked into the ground and recharged the water table. Thirty percent of the rain that fell on the forest ran off into other areas, 15 percent of it went into the atmosphere through the process of transpiration, and the final 5 percent evaporated into the atmosphere. The designers challenged themselves to match or improve on nature's ability to keep the water on site that falls on site. One of the strategies they are employing to accomplish this is to build bioswales on each corner of every intersection, each one being roughly the size of an ordinary parking space, that will collect every drop of water that would have gone into the city storm sewer. This water will percolate down through the strata of the swale and return cleaned to the water table. The design calls for shared-use treatment facilities for both gray and black water on site, which will further reduce the need for water moving from and to municipal treatment centers.

Lloyd's Crossing Ground Perspective.   *MITHUN, Seattle, WA. Reproduced by Permission.*

They are also working to create this densely populated area of the city as one that will also *live within its energy means*. About 20 percent of energy needed for the area will be supplied with photovoltaic panels. They will be incorporated into all south-facing building walls in cladding and shades, and arrays will be on rooftops alongside the green roofs. Wind power will generate the remaining energy needed through the local electric utility. It doesn't have to travel far because wind blows powerfully through the nearby Columbia Gorge. In addition to capturing the natural energies of the sun and wind, they're capturing the energy contained in the ground. A two- to four-foot wide geothermal, fluid-filled loop running underground through the district will share heating and cooling, as needed from building to building.

And contrary to a popularly held misconception about the cost of building in a way that will also support the Earth at almost every level, this turns out to be financially smart too. The design collaborators figured out how much it cost to send wastewater from the area to the nearest treatment plant, how much it cost to bring water into the area from the city water supply system, and how much it cost to supply the area with natural gas. It turns out this investment in infrastructure that enables the area to live within its energy means (potable water, black and gray water, solar, wind, and geothermal) will be completely paid back by the year 2030, the cash flow will be positive, and at the same time the city will be serving more citizens through the expansion of the area to 10-million square feet! And more than that, it will be living in the world in many ways as the conifer forest lived—it will make the air purer, add to the potable water available, and live on the energy given to it by the sun, the wind, and the geothermal Earth itself.

Lloyd's Crossing Aerial View.   *MITHUN, Seattle, WA. Reproduced by Permission.*

This is our best attempt so far to be generative designers in a place that houses huge masses of people. I applaud the attempt and believe that it succeeds on a number of levels. It does fall short on one level, and I bring it up not to criticize, but to help us get our heads wrapped around the difficult task that we face. In looking at a conifer forest as model, we need to identify what happens to a tree when it comes to the end of its natural life as *this* particular tree. It begins to decay, falls to the forest floor, and rots away. While it "lives" as a decaying tree, it releases nutrients back into the soil, providing fertilizer and mulch to nourish the Earth. It provides habitat and shelter for forest animals as well as a fecund foundation for lichens, moss, and so on. Even in death, it has a fruitful purpose. Another challenge we have is to create our buildings so that they have another fruitful purpose to fulfill when they come to the end of their life as *this* particular building. Now that we know that it's possible to live within our energy and water means, and to do so in such a way that it won't break the bank, it is time for us to look at this next challenge—generative design more fully applied to the built environment.

The word "generative" literally means "having the power to produce or originate." Its meaning when used in conjunction with the built environment implies that there is no waste, *and* that everything is a resource. Anything that is produced in the process of design and construction also has the power to produce or originate, and because of that, the end-life of all products and byproducts are valued and are designed so they are able to be used in a new way. On the other hand, degenerative in this context means that something is

1. Vertical Axis Wind Turbines
2. Vegetated Roof
3. Solar Shading
4. Cafe / Living Machine
5. Solar Control at South Facade
6. PV or SHW Panels
7. Rainwater Storage (opt.)
8. District Thermal Loop Connect to Building
9. To Subsurface Irrigation at Landscape Areas
10. District Thermal Loop Connect to Lloyd Center Tower
11. Catalyst Project

Lloyd's Crossing Section View.   *MITHUN, Seattle, WA. Reproduced by Permission.*

designed that has only one lifetime. When its life is complete, it serves no further purpose. It is waste and must be thrown away. It also implies that any of its byproducts are waste, as well—no thought has been put into the effect they may have on the Earth and its inhabitants.

Let's look at what this means in the actual state of interconnectedness of everything in existence that we've discussed in terms of both physics and ecology. Inside of that model, what would it mean to throw something away? The material item would be removed from our line of vision—thrown into the wastebasket, emptied into the garbage can, picked up by the garbage truck, carted away to the landfill, and dumped. Our trash nevertheless remains connected to everything in existence. As it lies in its "final resting place" in the dump, it begins to decompose and interact with the material that surrounds it. For some articles, that decomposition will take centuries, for others it will be swift. And depending on what material is decomposing next to it, our trash will produce something materially new that will range from harmless to toxic. But in any case, it won't be removed or disconnected from us—the gasses that the new material produces will be emitted into our atmosphere, into the very air we breathe. My point here is that the idea of discarding anything at all is centered in a logical fallacy. In this Earth system, situated inside of this universe system, where everything is connected, it is impossible to "throw something away." We can only place the discarded material outside of our own personal line of vision. That's all.

Let's examine what degenerative design in the built environment looks like inside of this model in which we understand that everything is connected. A very good example of degenerative design has happened in the building of single-family homes, using the same set of blueprints and specifications, which have been built throughout the United States, regardless of where they are located. Usually, the land gets leveled, regardless of what kinds of contours had existed on site originally. Streets are laid out to maximize the number of homes that can be built on the land. Then the "natural" features of the development get sculpted in to the subdivision. Because of all the impermeable surfaces being installed, little ponds are usually required as detention basins so that heavy rainfall doesn't inundate the storm sewers all at once. The homes are built with total disregard as to the natural energy that is available, and using structural and material guidelines that abide by the county's building code. (Building code, for those of you who don't know it, is the absolute minimum quality required in the construction of homes and buildings. So when a home or building gets "built to code," depending on the location and the county, it doesn't necessarily mean much in terms of quality.) These mass-built homes that are reproduced across the U.S. landscape rely completely on the infrastructure of gas and electricity to provide heating, cooling, and lighting because they don't take advantage of any of the natural energy available on site. They are good examples of degenerative design because they are built to have only one life. When we're done using any pieces or parts of them, they are discarded and hauled away to the landfill, and we don't give them another thought.

You might be thinking that it doesn't make any difference because the homes often last for several human lifetimes. But what if we designed our homes, office buildings, strip malls, and so on, so that they could be dismantled and reused when they've reached the end of their useful life in their current form? It may sound like an impossible proposition, but it isn't. We actually have historical examples of this kind of thinking. The picture below is one that I took at the Henry Ford Museum in Dearborn, Michigan.

It is called the Dymaxion House, designed by Buckminster Fuller in the late 1920s. The house was supported from a central column and was able to produce its own energy, recycle its own waste, and collect and store rainwater for use. It was designed to cost the buyer about as much as a Cadillac did at that time. It was to be manufactured, shipped anywhere in the world as a complete kit contained in a durable metal tube, and assembled on site by simply following the directions. There was no need to have utility infrastructure, making it very flexible as to where it could be set up. When a homeowner had to move, they could take the house apart, pack it up, and have it moved and reassembled at

Photograph Taken by the Author

the new location. The house is small by today's standards—just over a thousand square feet—and I'm not proposing that we re-create this package. But it worked as it was designed to work. And if a home could be conceived of and designed to do this in the late 1920s, then we can certainly accomplish this same kind of genius today in our built environment.

In nature on Earth, it's all generative design, and it all happens inside of a closed, finite, self-supporting system. When we look closely at what actually happens in our natural Earth world, we see that generative design is the underlying framework of what takes place. Everything generates. Everything has a next purpose in the system—and that purpose is meant to nourish, to foster fecundity and flourishing life. When we interject substances into this system, the system itself will continue to do what it does—to generate. This is what we have to be cognizant of when we build or create anything. We must make sure that when we interject a substance into the generative system that is nature, it can be used to benefit and nourish the system. When we interject a harmful substance into the system, it will not be set aside by nature. That harmful substance will be taken in and used to generate. Nature does not discern the difference between the two substances. In nature, we human beings are the representatives who have the ability to discern. We must begin to make decisions that fit within the framework of generative design, with next lives designed into the product, and produce only that which nourishes and fosters life.

Summary

In this chapter, we've discussed the difference between generative and degenerative design. "Generative" means "having the power to produce or originate." In the context of the built environment, it means that everything is treated as a resource, and the end-life of all products and byproducts are valued as such, and are designed so they are able to be used in a new way. "Degenerative" in this context, means that something is designed that has only one lifetime, and at the end of its life it is discarded as waste.

There is a new leadership emerging in construction-related fields that takes the generative design that occurs in nature, and incorporates it into the built environment. This thoughtful approach has design goals for projects that have them living within their water means and their energy means. These ideas are not new. One example given is of a house designed by Buckminster Fuller in the 1920s. He designed it so that it was able to produce its own energy, recycle its own waste, and harvest rainwater for use.

The way that nature works on Earth is all generative design, and it happens inside of a closed, finite, self-supporting system. Everything has a next purpose in the system—and the purpose is to nourish, and foster biological flourishing. When we interject substances into nature, nature will continue to generate—it does not discern the difference between harmful and beneficial substances—it simply uses what is there to generate. We are the ones who can discern the difference between harmful and beneficial substances. We are the ones who must begin to make decisions that fit within the framework of generative design that assists nature in its ability to support biological flourishing.

Questions and/or Assignments

1. Go to http://www.architecture2030.com and watch the "Face It" webcast. Write a couple of ideas about what you can do in your personal life to take the 2030 challenge yourself.
2. Write a couple of ideas describing how you might be able to take the 2030 challenge into the work you do.
3. Explain why it is important to stop using coal to provide energy for our buildings and homes.
4. Call your electric company and ask them if they are investing in any renewable resources to provide energy, or are they relying only on coal. Write their response.
5. Think about what you would need to do to "live within your energy means." Write out some actions that you can take right now. Write a plan for actions you can take within the next five years.
6. What do you throw away? Are you recycling everything that you can? Your assignment is to carry a trash bag with you for one week. Put in to the bag you are carrying everything you discard. When you have to carry your trash with you, do you make more of an effort to recycle? (The purpose of this exercise is to become conscious of waste.)

4

■ ■ ■

Whole Systems
Thinking

Abstract

Chapter 4 discusses the concept of whole systems thinking as it relates to the design and construction process. Traditionally, our design and construction process has reflected our tendency to treat individual parts as if they were unconnected, and this treatment causes problems. A generalized, brief version of conventional project organization is described, and an example of this kind of construction project is given to illustrate some of the pitfalls that can occur. Incorporating whole systems thinking into the process of design and construction can reduce or eliminate these pitfalls. The purpose of Occupational Safety and Health Administration (OSHA) is described. It is proposed that we may have to create a similar method in order to prevent companies from polluting, thereby compromising the health and safety of current and future generations.

Traditionally, the process of a building project, at every single stage, has been a very good mirror image of the view that Western philosophy and culture held—the one that portrays us as isolated individuals, independent and unconnected from each other. And as I mentioned before, this point of view is embedded in the way that we learn in school. We are taught to look at individual parts as the fundamental reality, and so it only makes sense that everything we do in life is subliminally invested with this perspective. But anything that we do, when it's based in this faulty way of conceptualizing the world, is going to create inefficiencies at every level. If individual parts are each designed separately by different specialists, and all of the parts are brought together to make up one complete whole, the best we can say about it is that the "whole" is comprised of a bunch of (however excellently conceived) things that are all glommed together. Some of the layers of inefficiencies that occur happen in the project organization itself, in relationships among participants, in the design process, and in the financial and environmental outcomes.

A generalized, brief version of conventional project organization looks something like the following. The owner hires an architect to design the building and specify the materials and also hires a general contractor to build the project. Most people who enter into this process want the most square footage they can get for the least amount of money, and that is often where the design attention goes. The general contractor puts the plans and specifications out for bids, and usually the lowest bidder gets the contract. If something goes wrong on the job and parties to the contract can't reach an agreement, they apply to mediation. If that doesn't work, lawyers are hired to resolve the issue for them. It is never a win–win situation, toward which Steven Covey's *Seven Habits* would have us strive. I am intimately familiar with that mode of job attainment, process, and resolution. As a primary owner of a commercial construction company, I've experienced plenty of projects that have worked like that.

I'd like to tell you about a job, in which we were the general contractor, that exemplifies this mode of operation. On this project, the public library board hired us to turn an empty, rundown ex-factory building, situated in the downtown area of a small, rural, Indiana town, into a library, senior citizens center, and childcare center. The building had some good structural bones that would remain. Located high on one of the masonry walls that would stay, there was a row of single-pane, original factory windows which had to be removed and the openings fitted with energy-efficient replacements. We ordered the new windows from the window schedule specification sheet in the blueprint package. When we began the window installation process, we discovered that the specified window sizes were incorrect. The architectural firm hadn't actually had anyone measure the openings. Instead of taking the normal amount of time to install, we had to do a lot of extra carpentry work to make the windows fit. For those of you who don't know it, that is a very expensive process.

I am proud of the work we did on this project—the building itself is a beautiful multi-use space. It is definitely made alive by its comfortable, bright places and by its layers of patterns of community use. But instead of making a small profit on the job as planned, it cost the company a lot to do the job. The architect wouldn't agree that we should have been able to rely on their specifications, and our only alternative would have been to file suit. We decided not to do that in this case. In other circumstances, we've made different decisions. In fact, you can't decide not to engage in litigation in cases like this very often, because you wouldn't be able to continue as a viable business. Yet, litigation is often the only option we're left with in this kind of construction model, and that makes it more expensive for everyone concerned.

Another problem with the conventional way in which we organize a building project is the isolation of the design process from the very people who will live and/or work in the space. When we design homes and buildings at the expense of, or indifference toward, the bodies and spirits of the people who will live or work in the space, something essential gets sacrificed. There is great need for the practice of architecture to evolve from being a work completed by one or more specialists to one that includes land, users, and all stakeholders in the creative process of design. This is not a new idea. In the 1970s Christopher

Alexander, Professor of Architecture at UC Berkley and head of the Center for Environmental Structure there, wrote and published a series of books that called for this transformation in the practice of architecture. For many of my friends who are architects, his books were required reading when they were in school. But the kind of change that Alexander calls for would be difficult to make in any profession. If we've been trained as "the expert" in any field, individually and culturally we tend to value the voice of the specialist more than we value the voices of untrained folk.

Although some architects have transformed their practice, there are many who have not. Last year a couple came to our construction company because they were interested in hearing more about our emphasis on building sustainably. They told us that they had originally hired a well-known architect who is also a professor of architecture at a university. The man never asked them any questions about what they wanted to do when they lived in the home they wanted to build. He never went to look at the land. He didn't consider site orientation or how to capture any of the natural energies available. He simply produced a design and a set of prints that had his "signature look" but which was wholly unconnected to any of their desired patterns of use. (This is an extreme case, but as improbable as it sounds, variations of this same attitude still occur.) So they began searching for someone who would include ideas about the way they wanted to live in their home and on their land. They wanted ideas that would include the natural energies available, especially in terms of winter sun for passive warming and summer breezes to augment cooling. They heard about our company focus on sustainable construction and that is how they ended up working with us.

Our architect asked them how they wanted to live in the house, and about what was important to them. What kind of spaces did they want and need? This was to be their retirement home. He is a bread maker by avocation, and does so most weekends. They both enjoy cooking and envisioned an herb garden and bread bakery that would evolve into a small business that would supply local restaurants when they retire. One of the most important features in their home became the professional-level kitchen with a built-in clay bread oven. Another feature they wanted includes a window wall that opens onto a screened porch just off their bedroom. On warm summer evenings, they open the window wall and simply roll their bed onto the porch for sleeping in the cool night air. Communication is one of the most important activities in the construction process.

The way in which the conventional building process has worked makes it both financially and environmentally expensive. Financially, some of the folks who lose at this game are the people who end up paying for the costs to operate the building over the long term—the owner or tenant. If we're talking about any building that is built with taxpayer's money—like a school building, or a city, state, or federal building, then the operating expenses come from the taxes that you and I pay, and we're the ones for whom it is financially expensive. If the building is a long-term user like a hospital or church, we're still the ones who carry the extra operating expense through our insurance premiums and tithes.

If we try to construct a sustainable building using the conventional process that treats systems as separate and unconnected, we end up just adding "green" items to the project, which adds more expense. That action has caused the idea to proliferate that building sustainably is too expensive for the average person to bear. If we design to ensure that passive heating and cooling is available in a building, and we use materials that have an especially high insulation value, and at the same time, we install a heating, ventilating, and air conditioning (HVAC) system that doesn't take any of those things into consideration, then we're spending more money than is necessary. That's a very simple example, but it is exactly the kind of outcome that happens when we don't treat a project as a whole system, embedded in the interdependent systems of which our planet Earth is made.

The process itself must change. Just as the old method of operating mirrored the linear thinking of the industrial age, the new process that we'll describe here mirrors a more holistic, creative way of thinking. We begin with an integrated design charrette. In this new way of operating, *we work toward reaching a set of goals by optimizing relationships—in people and systems.* We start by setting the goals that we want to achieve and gathering as many of the "systems representatives" at the table as possible. Depending upon the size of the structure, we could have the owners of the project, the architect, builder, mechanical, electrical, and structural engineers, landscape architect, representatives of all subcontractors, waste and water specialists, perhaps an energy modeler, a commissioning agent, and so on. If it's a home, the homeowners will be at the table; if it's a building project, folks who will be working in the building should be present in addition to the owner. A friend of mine who recently completed an integrated design charrette mentioned some good-humored rules of thumb. He said that it's beneficial to have everyone check their egos at the door, and to have a good therapist at the table! He was only half kidding. When we're used to dealing in a top-down, didactic way as is the case in a conventional building process, we have a considerable internal and relational adjustment to make in order to work collaboratively with others toward a goal.

There are other adjustments we'll have to make as well. Our society has allowed people to make money at the expense of the rest of the world since the industrial revolution began. We've had to create gigantic systems like OSHA in order to protect the people who work on the job, making sure that safety measures are taken. Despite the rules put in place, and the fines levied against companies, people are still killed on jobs every year because of inadequate or nonexistent safety measures and equipment, or because the worker doesn't follow the safety guidelines. We recognize the need to have the OSHA guidelines to protect company employees. What is not yet recognized is that the pollution in our environment is compromising the health and well-being of current and future generations. The construction business culture is still very tolerant of an owner who voiced the opinion about green building during a recent contractor's association board meeting, stating that all he cares about is "the green that I can rub between my fingers." Some of the other owners at the table chuckled. I bring this comment up to help us realize that lamenting about the attitude of the person who would make that statement and/or chuckle at it isn't going to help. But managing the process differently will help.

In the design charrette, the first order of business is to set goals for the project as a whole system. In order to do that, we strive to address as many of the following questions as possible. What are the natural ecosystems in which this building will be participating? What can we do to make sure that this building both contributes to and benefits from those natural systems? What was the carbon footprint of the land before human beings developed it? What existed on the land? What flourished and why? How much water did the land absorb? How much sunlight and carbon dioxide did it take in? How much oxygen did it generate? How can we ensure that the new structure will live within its water means? How can the new structure live within its energy means? What will the next life of the structure be? How many functions can the new structure serve? Will the materials it generates produce the nutrients necessary for this particular piece of land to flourish?

There is one more item that I'd like to add to the mix here before we move on to the process and description of the design charrette in the next chapter. I would like you to think about *how you feel* inside of the buildings in which you live and work. Are there spaces that make you feel fidgety and uncomfortable when you must stay in them overlong? I ask you to think about those spaces. See if you can identify what it is about the room that makes you feel that way? Notice also what it is about spaces in which you feel comfortable that invoke that feeling in you. There are hundreds of patterns that we human beings have always used in our buildings to make the built environment itself *inspire* the living that is meant to take place in the created space. What doesn't often get considered is that in order for spaces to inspire, the folks who are going to use them need to give input into the design itself. The aliveness of a space is caused by use, and it is used because it holds a central quality that is generated by the living patterns of use that take place in the space. The more patterns of use it has, the more alive it is. "Like ocean waves, or blades of grass, its parts are governed by the endless play of repetition and variety created in the presence of the fact that all things pass. This is the quality (of aliveness) itself."[1]

Summary

In this chapter we've discussed the concept of whole systems thinking as it relates to the design and construction process. Traditionally, our design and construction process has reflected our tendency to treat individual parts as if they were unconnected, and this treatment causes problems. A generalized, brief version of conventional project organization is described, and an example of this kind of construction project is given to illustrate some of the pitfalls that can occur. The conventional building process can be both financially and environmentally costly. When we change the process to one that treats it as a whole system, we prevent ourselves from making these financial and environmental mistakes.

The process itself must change. Incorporating whole systems thinking into the process of design and construction means that we begin with an integrated design charrette. We begin by working toward a set of goals by optimizing relationships in both people and systems. We must make other adjustments as well. OSHA was created to make sure that companies had proper safety equipment and practices in place for workers. Most companies did not do what was needed in order to protect the health and safety of workers without the urging of OSHA. We have a similar condition in terms of the environment. Companies are polluting our environment and, in doing so, are compromising the health and safety of current and future generations.

Questions and/or Assignments

1. What are the differences in the design process between a conventional and sustainable building process?
2. The new wisdom in building design is to include building occupants in the design process. Why is that important?
3. In what ways (and to whom) can the conventional approach to building be more expensive than the sustainable approach?
4. Why was OSHA established? What is its purpose?
5. What is an integrated design charrette, and who should attend?
6. What is a "carbon footprint"? What does that mean, and what is *your* carbon footprint?
7. Think of a room or building that inspires the patterns of activities that are supposed to take place in it. Describe it.

Notes

1. Thomas Alexander. *The Timeless Way of Building.* New York: Oxford, 1979, pp. x–xi.

5

■ ■ ■

Collaboration as Sustainability in Action

Abstract

Every sustainable construction project begins with an integrated design charrette. Chapter 5 reveals the individual and group characteristics needed in order to create a successful charrette. It is critical that all participants in a design charrette are willing to learn and that they have the ability to think in terms of whole systems. Other characteristics that people must have in order to function together well as a team are the individual abilities of *aspiration* and *personal mastery* and the ability to *understand complexity*, to *think in terms of whole systems*, and to *have reflective conversation using both mental models and dialogue*. This chapter briefly describes each one of these. We follow the first building project to achieve Leadership in Energy and Environmental Design (LEED®)[1] Platinum Certification in Indiana, Merry Lea Environmental Center of Goshen College, to illustrate the process of a design charrette.

Every sustainable construction project begins with an integrated design charrette. In this chapter I strive to articulate what it takes in terms of individual and group characteristics to create a successful charrette. A charrette is a collective brainstorming session with as many of the stakeholders participating as possible. I'll also describe a recently built project and the design charrette the group underwent in order to accomplish the first buildings in Indiana to achieve Platinum level LEED certification. The Merry Lea Environmental Learning Center of Goshen College is a 1,150 acre land and nature preserve, located in rich farmland close to Wolf Lake, Indiana, at US Hwy 33 and Indiana Hwy 109. The land preserve contains several lakes, bogs, and wetlands as well as forests, prairies, and meadows. They have recently completed phase 1 of a building project for collegiate student housing and an education center. The Executive Director of Merry Lea, Luke Gascho, is responsible for the overall environmental integrity of the project.

Rieth Village, Merry Lea Environmental Center, Goshen College, Wolf Lake, IN.   *Photograph Taken by Dr. Luke Gascho. Reproduced by Permission.*

In preparation for the design process, Luke traveled the United States to review a number of sustainable design projects and to conduct interviews with the owners to find out what strategies worked and those that did not. He visited the Environmental Studies building at Oberlin College, the sustainable housing and classrooms at California Polytechnic (John T. Lyle Center), The National Renewable Energy Labs in Golden, Colorado, and the Rocky Mountain Institute (RMI) in Snow Mass, Colorado. During the trip to RMI, he discovered that the staff there could be hired to lead the design charrette. Engaging them gave him the confidence to go forward, knowing the team would have this kind of expert help to guide them in working through the sustainable and ecological issues. They determined that using this charrette process would cost more money in the design phase, but would greatly improve the buildings' performance. This "front loading" of the design process has been shown to save both energy and costs in the construction and operation phases.

In 2001, charrette participants gathered on the Merry Lea site. A successful design charrette requires that representatives of all disciplines involved in the design build process be present together with representatives of the future occupants of the completed building. For this charrette, there were six to eight Merry Lea staff members present, two Goshen College students and three administrators, three Merry Lea board members, and representatives from an architectural firm, an engineering firm, and a construction company. The Building Commissioner and Building Inspector from the County Building

Commission also attended the educational parts of the charrette, which helped them understand some aspects of green building with which they were unfamiliar.

In a design charrette, it is critical that *everyone* at the table *be willing to learn*, and to *think in terms of whole systems*. Participants must be able to conceptualize the integrated nature of the whole. It is critical in this kind of integrated design process that those involved are able to listen to each other and to keep the dialogic process open and flowing. The consequence of stopping the process of dialogue is just as severe in nature as it is in our relationships. When we stop the dialogue—stop listening to and speaking with nature—that fact reverberates through nature's ecosystems and affects everything else. We have stark evidence of this on a grand scale in the situation we are currently facing with the issue of global warming. The Earth and its ecosystems are speaking to us very loudly right now. Whether we are listening or not is still an open question. The scientific evidence is startling, and the scientific community is in total agreement with the fact that global warming is taking place at levels that have never occurred before in the history of the Earth, yet the average guy on the street doesn't know this because so many journalists, in mistakenly striving to give their conception of an unbiased opinion, raise a question about the believability of the evidence. I hope that situation is different by the time you are reading this book, and that the voice of the scientific community is able to be heard and accurately reported.

The ability to *conceptualize the whole*, to *listen both deeply and openly* (in effect, transactively), and to *have a willingness to learn* is so essential in the design charrette process that you will want to choose participants for the team that have these characteristics. Teams must have certain core learning abilities in order to function well together *as a team*, consisting of both individual and group abilities. Peter Senge does a good job of identifying these individual and group abilities in his book *The Fifth Discipline: The Art & Practice of the Learning Organization*. They are reviewed as follows. However, if you would like to participate in, or conduct, a design charrette yourself, I'd suggest you read further. According to Senge the necessary individual abilities are as follows: *aspiration*, *personal mastery*, and the ability to *understand complexity*, to *think in terms of whole systems*, and to *have reflective conversation using both mental models and dialogue*. They are summarized briefly below to give you the means to understand them.

Aspiration is the ability to truly care about the project. The people at the table need to believe that it is an important endeavor to build sustainably. They don't need to know how to do that yet, but they need to aspire to learn how. The concept of aspiration is easy to understand. Just remember the last time you participated in a team activity—it can be something as mundane as conversing at the family dinner table. If one person is busy multitasking, either text messaging to someone who isn't present or trying to catch the score of the game on television, they are unable to participate in the conversation. The feeling they give to the rest of the "team" is that they don't truly care about it because their attention is divided. It may be that they don't understand how important conversation is in the art of maintaining relationships, and so, they divide their attention. We can also look to sports for examples of aspiration. None of the amazing successful passes, the almost miraculous three-point shots that make it, the nearly super-human push that athletes exhibit when they near the end of the course—the Lance Armstrong's and the Dave Wattle's of the world—would have occurred without aspiration. If you've ever watched the faces of runners approaching the finish line after 26 miles of running a marathon, or done this yourself, you know what aspiration is. If you are reading this book as part of a college course, you know what aspiration is from the inside out. We all aspire to achieve and/or to learn when we participate in a university course of study, aiming for a diploma.

Personal mastery is another individual learning ability a person must have. It is the ability to hold both an internal personal vision and an accurate view of the current reality *at the same time* and to have a sense of shared vision with the team. Personal mastery is a characteristic of a person who is committed to lifelong learning of a specific kind. It "is the discipline of continually clarifying and deepening our personal vision, of focusing our energies, of developing patience, and of seeing reality objectively."[2] When we have personal mastery, we know what really matters to us because we examine our values often enough to live our lives "in the service of our highest aspirations."[3] When we have personal mastery, we have a vision for ourselves which exemplifies our highest values, *and* we can see the difference between the vision and current reality. This is not comfortable. It requires that we hold strongly to the vision without adjusting it downward to get it closer to the existing current reality. It also requires that we don't deceive ourselves by altering current reality in our minds, moving it closer to the vision so that we can feel more comfortable internally. Achieving the ability to live with a sense of personal mastery means that we approach our lives from the internal viewpoint that *we create our life*. The contrasting view is one that is more common. It is the perspective that is primarily reactive and views life as something that happens to us—whether the experience is one that we see as being good or bad.

Mental models are the "deeply ingrained assumptions, generalizations, or even pictures or images that influence how we understand the world and how we take action."[3] They are usually under the radar of our conscious awareness, yet they greatly affect how we behave in the world. Some of them are easy for us to identify when we begin looking for them. They are the origins of our prejudicial thoughts, feelings, and behaviors. Others are much more difficult to spot. An example of an insidious mental model that I held at one time is the idea that *girls are not good at math*. Growing up, I believed that I wasn't good at math, and that is exactly what got reflected back to me in terms of what I was able to learn and the resultant grades that belief generated. I put off going to college because I didn't really believe that I was smart enough to do it. When I was 30, I went to college and discovered that I loved math! I earned A's in all math classes, and eventually earned a doctoral degree in Philosophy which required advanced levels of math and logic classes.

Mental models affect our ability to work well with others as a participant in a design charrette. Because they are often unconscious, as was my earlier belief about my math ability, they are extremely pernicious. One that we can identify as prevalent in our culture and that also prevents us from making real progress relationally and dialogically is "the need to be right." If we need to be right it means that we will be invested in our own ideas to such a degree that we only advocate for them. That need to be right actually renders us incapable of listening to another person openly enough to actually consider and value their contribution. In order for an integrated design charrette to function well, the group has to be able to become a learning team together. And how well are we able to learn if we just think that we've got it right and others have it wrong?

Another core learning ability that teams need to have is the ability to *dialogue*. For team learning to occur, dialogue is a component of *reflective conversation*. Team learning starts with dialogue, defined as the ability of individual team members "to suspend assumptions and enter into a genuine 'thinking together.'"[4] In our culture, learning has come to mean the simple act of taking information in, but that's only a small part of what learning actually is. When we learn something, it changes us. We become able to do something that we've never been able to do before, or we are able to think about something in a completely new way. In effect, when we actually learn, we "extend our capacity to create, to be part of the generative process of life."[5]

The Merry Lea integrated design team successfully became a learning organization, able to think together to accomplish the goals of the charrette. They began by discussing the interpersonal skills each needed to have in order to become a team able to think together. (One original member of the design team exhibited such a strong mental model of "the need to be right" that he had to be replaced on the team.) Participants didn't necessarily know a lot about green building, so the morning of the first day of the charrette was devoted to education about different principles of sustainable design. Luke explained the goals of the charrette, followed by a variety of presentations about green building case studies, energy approaches, lighting principles, site and water design, and alternative wastewater treatment options. In the afternoon they established work groups to explore possible design ideas. One group thought about design concerns addressing issues of progressive alternative wastewater and storm water treatment systems, pervious parking and trail designs, site erosion controls, building orientation, and restorative native landscaping. Another group talked about wanting the buildings themselves to be congruent with the material that will be taught in them. They discussed mechanical and lighting goals and systems, building envelope, and interiors. The facilitators from RMI helped the group to prioritize green-building objectives and to evaluate their choices in light of the United States Green Building Council (USGBC) LEED standards. It was at this point that the group realized that working toward a Platinum LEED rating was a realistic goal. The set of objectives selected at the charrette informed choices of the integrated design team in the months of planning that followed.

The minute you enter the Merry Lea property, you can see that respectful and thoughtful planning has gone into this work. The evidence is palpable that the charrette team was able to become a learning organization.

A summary of some of the sustainable features that were designed into the project follows. The buildings are oriented and designed so they can utilize passive solar heating and day lighting, and photovoltaic solar panels that produce electricity, which meant that size and energy loads of heating, ventilating, and air conditioning (HVAC) units are reduced.

Gravel Grass Parking Area at Entrance to Rieth Village at Merry Lea.   *Photograph by Dr. Luke Gascho. Reproduced by Permission.*

A

B

(A) Passive Solar Heating and Day Lighting. (B) Solar Hot Water Panels.   *Photographs Taken by Dr. Luke Gascho. Reproduced by Permission.*

Geothermal heat pumps augment the passive heating and cooling. The project has triple pane, operable windows to provide passive ventilation on warmer days. A 4-kilowatt array of photovoltaic panels produces 18 percent of the energy needed to operate the buildings. A wind turbine also adds to the renewable energy fund, providing about 16 percent of the energy needed by the complex. In the landscaping, they've eliminated some invasive species and have incorporated native plants, which reduce stormwater runoff and eliminate the need to provide additional water or fertilizer for plant maintenance. In addition, strategies were employed to prevent stormwater runoff. The parking bays are finished with gravel grass, sidewalks are constructed of porous concrete, a cistern is installed to recycle rainwater, and two rain gardens have been created. The gray water from the cistern is plumbed to the toilets so that potable water doesn't have to be used for flushing. The waste water and sewage is treated on site in a treatment system that utilizes constructed wetlands to do the job.

You've now got a decent description of the personal attributes that folks need to have in order for a successful charrette to occur. You also have a description of the order in which events took place in this charrette for Merry Lea Environmental Center. And finally, you've got a brief overview of the outcome of the charrette. The most difficult subject to describe is the process of the charrette itself. What has to be created is an experience that has a minimal amount of structure so that the creative work of brainstorming can occur. This can be felt uncomfortably like chaos to participants. The quality of the final

Potable water system on the left comes from the well. Ground source heat pump is in the Center. Nonpotable water system on the right comes from the cistern. *Photograph Taken by Dr. Luke Gascho. Reproduced by Permission.*

Wastewater Treatment Wetland Cell Being Planted.   *Photograph Taken by Dr. Luke Gascho.*
Reproduced by Permission.

outcome will depend on the creativity that participants are able to achieve, and creativity is stifled by structured outcomes.

If we're going to achieve out-of-the-box thinking and a genuinely fruitful charrette, we must provide only the bare-bones structure of the charrette session. Identify when breaks for food and drink will be available. Start with an educational process so that everyone present understands the general kinds of problems that sustainable construction addresses. Look at the site, including everything that surrounds it, and identify its attributes and defects. If the owner has not set the goals to be achieved in terms of energy efficiencies, set the goals with the group as a whole. Begin the process of brainstorming solutions that will achieve the energy efficiency goals that have been established. This is the part that feels like chaos; resist the urge to control it. I always supply tables with a large aerial view of the site, a survey with trees identified, tracing paper, large sheets of drawing paper, and a supply of colored markers. When I'm able to restrict myself to supplying only the bare bones of a structure, I am always amazed at the originality of the outcome.

Summary

Every sustainable construction project begins with an integrated design charrette. This chapter reveals the individual and group characteristics needed in order to create a successful charrette. It also describes the first buildings to achieve LEED Platinum Certification in Indiana, the recently constructed Merry Lea Environmental Center of Goshen College, to illustrate the charrette process.

It is critical that all participants in a design charrette are willing to learn and that they have the ability to think in terms of whole systems. In addition to these core learning abilities, people must be able to listen both deeply and openly. Peter Senge identifies the characteristics that people must have in order to function together well as a team. These individual abilities are *aspiration*, *personal mastery*, and the ability to *understand complexity*, to *think in terms of whole systems*, and to have *reflective conversation using both mental models and dialogue*. This chapter gave a brief explanation of each one of these.

Many of the sustainable features that were designed into the Merry Lea Environmental Center are also described in this chapter.

Questions and/or Assignments

1. The title of this chapter is "Collaboration as Sustainability in Action." Explain what that title means. Why is collaboration an example of sustainability in action?
2. What is a "whole system"? Give an example.
3. Why must every participant in a design charrette be willing to learn and to think in terms of whole systems?
4. What is personal mastery, and why is it necessary for participants in a design charrette to possess this quality?
5. Do you possess the quality of personal mastery? Give an example from your own experience of either yourself or someone you know exhibiting this quality.
6. What is "reflective conversation" and why do members of a design charrette need to have the ability to do this?
7. The author states that team members of a design charrette must have the ability to enter into the experience of being able to suspend assumptions and think together. Why is this ability important?

Notes

1. LEED is a rating system developed by the United States Green Building Council to establish criteria that regulate and give definition to the term "green construction." You can get more information by going to their website: http://www.usgbc.org.
2. Senge, Peter M. *The Fifth Discipline: The Art & Practice of the Learning Organization*. New York: Doubleday, [1990] 2006, p. 7.
3. Senge. *The Fifth Discipline: The Art & Practice of the Learning Organization*. p. 8.
4. Senge. *The Fifth Discipline: The Art & Practice of the Learning Organization*. p. 10.
5. Senge. *The Fifth Discipline: The Art & Practice of the Learning Organization*. p. 14.

Sustainable Construction Roadmap

6

■ ■ ■

Site and Natural
Energy Mapping

Abstract

This chapter describes the conventional approach to building regarding the way the site itself is treated, and explains the different approach taken if we consider the site with sustainability in mind. The chapter also gives the steps that need to be taken to collect data about the site, and includes site research components, directions for conducting a site visit, and directions for energy mapping and modeling. This chapter also introduces a commercial project that we will follow through its construction and submittal process as it seeks LEED NC Gold certification. The chapter explains LEED Sustainable Sites by following the project through each prerequisite and credit, and includes the LEED scorecard and cost analysis the LEED-AP and owners used during this project.

In the conventional approach to building, the site is cleared and leveled. It is the most common way of preparing land for construction. During the last couple of years, two plots of land near my house that had been densely forested with a variety of old growth trees, most ranging in height from 30 to 60 feet and taller, were completely cleared to make the sites ready for building in the conventional way. One of those plots now contains a dozen apartment buildings; the other is undeveloped, waiting to be sold since the trees were removed and the land was leveled. The land with the apartment buildings being built on it has no shade trees left with the exception of some located at the rear of the property along a drainage ditch. Some very small front and back yards of traditional grass have been planted, and bits of shrubbery have been introduced along the road and planted against the fronts of buildings in the complex.

In a green building approach, it is extremely important that we look at environmental issues contextually throughout the process. So when we consider the site, first we look at the total context in which it exists. The Earth is comprised of a number of "biomes"—large areas consisting of similar vegetation, animals, and micro-organisms. Biomes are classifications of large ecosystems, for example, mountains, deserts, rainforests, savannah, the northern conifer forests, and the midwestern plains. Here in the midwest, our biome was originally made up of temperate forested areas and plains of prairie grasses. Some of the forested areas remain, as do some of the plains; but mostly we now have a mixture of farmland cultivated and planted with hybrid corn, soybeans, and wheat, and our spreading cities and suburbs. Wherever the construction site is located, figure out what the pre-existing biome was. The interaction of the sun with the ecosystems that work together inside of all of the Earth's biomes creates the climate that makes our planet inhabitable.

The interdependency of organisms is the ecosystem within which any site exists. An ecosystem is the dynamic way that living things in nature reciprocally interact with each other in a symbiotic relationship, creating the conditions for survival. In order for this reciprocal interaction to occur, the ecosystems must remain in balance. Ecosystems can be as large as the Sahara desert, or as small as a little pond. When we make any alterations to a site, we have to determine how we can maintain the balance in the ecosystem—its carbon cycle, nitrogen cycle, water cycle, and food chain.

Looking at the land in my neighborhood that has been changed from an old-growth arboreal forest to apartment complex, we can see what has happened

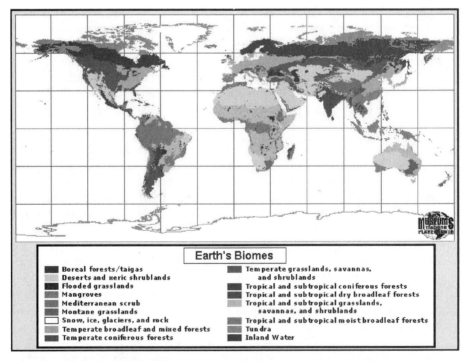

Earth's biomes.   From http://earth.rice.edu/mtpe/bio/biosphere/topics/biomes/biomes_map_final.gif

to the ecosystem involved in this site. The trees took in carbon dioxide, "exhaled" oxygen, and used energy provided by the sun and the Earth's nutrients to flourish and produce food for other organisms. The apartment complex is fitted out with conventional furnaces and air conditioners, and has code-level (the minimum required) batt insulation. It disrupts the carbon cycle by emitting more carbon into the atmosphere. *In order to build sustainably, we would need to find a way for the apartment complex to take in carbon dioxide and exhale oxygen, just as the pre-existing forest had, or be carbon-neutral.* We'll address this in Chapter 7.

The nitrogen cycle of the former forest, now apartment complex has also been radically changed. The brief way to describe this change is that because the apartment complex uses fossil fuels for heating, cooling, and electricity, nitric oxide is added to the atmosphere. When nitric oxide combines with other elements existing in the air, it eventually creates acid rain, which damages living things. The plants and trees that were in the forest metabolized the nitrogen from the soil and turned it into protein that was then available for the food chain in the ecosystem. *In order to build sustainably, we would need to find a way for the systems of the apartment complex to metabolize nitrogen and turn it into protein, just as the forest had.* And yes, we can do this. We'll also discuss this in Chapter 7.

The water cycle of the former forest has also been radically changed. In the forest, much of the water that fell on the land stayed there. The soil and plant material acted like a sponge, with all of its roots either absorbing or retaining water. The roots cut deep channels into the clay soil and create a pathway through which the water can flow back down to the water table. The apartment complex, filled with hard, impervious surfaces, in the roofs and pavement, funnels the water away from the land, into a ditch and the city storm sewers which drain to the rivers. Uncontrolled runoff can subject a region to "100-year flooding" at regular (short) intervals. *In order to build sustainably, we would need to find a way for the apartment complex to keep the water on the land that falls there, and allow it to find its way to the water table, just as the forest had.* We'll discuss methods to accomplish this in Chapter 8.

The nutrient chain of the former forest has also been radically compromised in the construction of the apartment complex. The nutrient chain has a number of components. The sun is the source that provides energy for every living organism. Plants transform the energy of the sun to make food for themselves and for other participants in the nutrient chain. The next level consists of the plant-eating animals (herbivores), and the following level consists of the animals that eat the plant-eating animals. The final level of the food chain is the living organisms that "eat and digest" dead organic matter, helping it to decompose and become nutrients in the earth once again. When we plant things that require pesticides and herbicides, as the lawns and shrubs do of the apartment complex, we're compromising the food chain (which as herbivores and/or carnivores, we are part of and must have in order to survive). *In order to build sustainably, we would need to adopt methods for the apartment complex to contribute in a positive way to the nutrient chain, as the forest had.* Chapter 7 will address some methods for accomplishing this.

Building sustainably requires that we understand that *whatever we do will alter the ecosystem* within which the site functions, and so we must make sure that we restore balance to the ecosystem by adding nutrients to the earth, restoring more of the water to the water table, and making the air as clean as or cleaner than it was as a forest. With that in mind as a primary goal, our first task is to conduct site research and map the natural energy systems on the site. Some of these steps are part of the conventional method of design and construction, so you'll recognize them; but the difference is in what solutions are derived from the information gathered.

1. Data Collection, Site Research
   - Wetland search: Go to the U.S. Fish and Wildlife Services online wetland mapping service to ascertain whether or not there are any wetland areas on the site. The website address is http://www.fws.gov. Click on "wetlands," then on "wetlands mapper quick launch." A map of the United States will come onto your screen and you can zoom into the area of your site.

     This is an example of the kind of image you'll get when you conduct a wetlands search. The image below portrays the wetlands in my neighborhood and also shows the apartment complex site, denuded of trees and all foliage at the top of the photo.
   - Floodplain search: Go to the FEMA website to find floodplain mapping at http://www.msc.fema.gov.

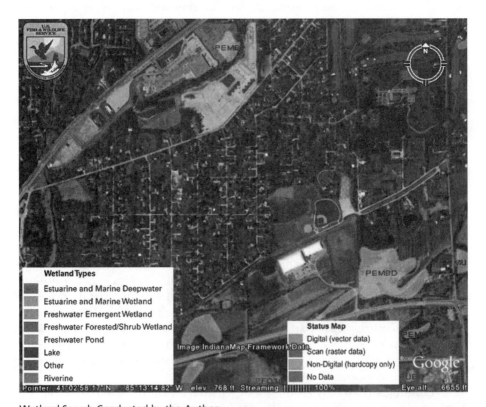

Wetland Search Conducted by the Author

- Aerial photos: In Indiana you can obtain aerial photos at http://www.indiana.edu/~gisdata, but do consider when the photo was taken as something could have happened to alter the land in the meantime. This is a site sponsored by Indiana University. Counties often have aerial photos available.
- Deed research: Look for past owners and uses of the site and surrounding area for any problems that may have occurred to contaminate soil and/or groundwater, "brown field sites" are one example.

2. Data Collection, Site Visit
   - Go to the site, walk it, and make note of the following site conditions.
   - Identify existing trees, their size, and other vegetation present. Note which are indigenous to the area, and if there are any invasive species present. If you are unable to identify plants and trees yourself, you can call your local county agricultural office, or ask a botany instructor at a local high school or college for assistance.
   - Interview neighbors if you can. They will often be able to tell you about seasonal issues, or things they've observed that affect the property, but that wouldn't come to your attention in the normal course of a project. On the Habitat for Humanity sustainable residential project that you'll read about in Chapter 13, conversations like this brought to light the fact that an area on the southwest side of the lot flooded whenever it rained for several days in a row. They also told us about what they knew had been dumped on the site. Evidently, some years back, county trucks had backed into the site and dumped chunks of concrete. The site had also been the recipient of a number of used tires. This information helps us plan for extra site preparation and/or reparation costs.
   - Complete a topographical survey, and map the location of indigenous trees. You may want to clear the invasive species before you do this. The reason it is so important to use indigenous trees and plants is that the root system is specific to the soil. In our region's hard clay soil, indigenous plants have developed root systems that grow strong and deep. They enhance the ability of the clay to allow water to move down through it, and also grow deep enough to reach available water and nutrients. Once established, indigenous plants will thrive through all of the extremes of the local climate.
   - Also make note of the area surrounding the site. Note any buildings or trees that you will want to consider in the process of planning access to heating, cooling, and day lighting using the natural energy available on the site.
   - Photo documentation: Stand at the center of the site and while turning 360°, snap photos that you can piece together into a radial view of the land. Take pictures as well of anything surrounding the site that may need to be considered as you work on the design.

3. Data Collection, Energy Mapping and Modeling: Climate data can be found using the free, downloadable software called "Climate Consultant." It is available at the following website: http://www.eere.energy.gov/buildings/tools_directory/software.

- Wind: Find information about prevailing winds, average speeds, and a plethora of other information for the United States as a whole and for your specific location by going to http://www.eere.energy.gov. Specific maps are located on the website at http://www.eere.energy.gov/windandhydro/windpoweringamerica/pdfs/wind_maps/us_windmap.pdf. Map how the wind moves across the site in both summer and winter.
- Solar: Find information about degree heating and cooling days from *The Passive Solar Energy Book* by Edward Mazria.[1] This book is the seminal "how to" book to assist in the process of incorporating passive solar into a building. In it Mazria explains how to calculate how much heat you can count on receiving from the sun, at your latitude, on the twentieth day of any given month, in Btu's. Using a solar calculator, map how the sun moves across the site in both summer and winter. There are also web resources available.
- Use available energy modeling software. The U.S. Department of Energy has an energy modeling tool that you can download for free. It models heating, cooling, daylighting, and other energy flows in buildings. You can use this tool by inputting text; the software reads the information and gives the energy modeling back to you in text. There are graphic tools that interface with this free tool, so if you want to see the modeling in three dimensions, you can do that too. This program is available online at http://www.eere.energy.gov/buildings/energyplus/. Another free whole building energy analysis software that is available from the Environmental Protection Agency (EPA) is called "The Green Building Studio." It is a good program and is easy to use.

If you have any input into choosing the location of the site, do what you can to ensure that it is location efficient. In the conventional approach to building, the transportation is assumed to be the individual car, and location efficiency is not considered. If you can influence the decision, help to choose a site at which infrastructure is already present and mass transportation is easily accessible. That is, the one sustainable aspect the apartment complex I've talked about has is that mass transportation is available at the corner, and mixed use planning where occupants can walk to the nearby grocery store, eateries, and shopping center.

USGBC LEED NC CERTIFICATION PROCESS

Case Example: Sweetwater Sound, Built to LEED NC Gold Certification Level

In order to help explain and illustrate the LEED certification process, we will follow the construction and submittal process of the recently completed headquarters for Sweetwater Sound, a music instrument and recording retailer. Their facility is a 155,300-square-foot building consisting of an office area,

One of Several Recording Studios in Sweetwater Sound.   *Sweetwater Sound/John Hopkins. Reproduced by permission.*

warehouse, and auditorium and recording studios. They have worked toward achieving a LEED Gold certification, attaining the following LEED credits:

Sustainable Sites	7 (out of 14 possible points)
Water Efficiency	5 (out of 5 possible points)
Energy and Atmosphere	8–13 (out of 17 possible points)
Materials and Resources	7 (out of 13 possible points)
Indoor Environmental Quality	12–14 (out of 15 possible points)
Innovation in Design	5 (out of 5 possible points)
Total Points	44 to 51 (out of 69 possible points)

If they achieve the maximum they are expecting, 51 points, they are likely to make decisions to go forward with some of the things they are considering in additional energy efficiencies. Attaining just one additional point will take them into the platinum range of LEED certification. I've chosen Sweetwater Sound to use as a case study because I think the way they've made decisions illustrates the real-world process of a LEED® construction project well.

The levels of LEED certification are as follows:

Certified:	26–32 points
Silver:	33–38 points
Gold:	39–51 points
Platinum:	52–69 points

USGBC LEED NC: SUSTAINABLE SITES (14 POSSIBLE POINTS)

Prerequisite 1: Construction Activity Pollution Prevention (Required)

The intention of this prerequisite is to control erosion, sedimentation, and dust from construction activities. To comply with this requirement, an Erosion and Sedimentation Control (ESC) plan must be created and implemented. The plan should comply with the 2003 EPA Construction General Permit or local codes—whichever of the two is more stringent. The ESC plan should identify what you are doing to prevent soil loss, prevent sedimentation from moving into streambeds or storm sewers, and prevent air pollution from occurring due to particulate matter being released during construction. This prerequisite is required for all projects regardless of site size. Develop the ESC plan during the design phase of the process, very early on. Some strategies you might use are things like mulching, seeding, creating earth dikes and sediment traps, and erecting silt fencing.

All projects seeking LEED certification of any level must perform all prerequisites in every area. Sweetwater Sound accomplished this prerequisite by creating and implementing an ESC plan. They used silt fences to prevent soil from running off of the land.

Silt fencing at Sweetwater Sound.   *Sweetwater Sound/John Hopkins. Reproduced by permission.*

SS Credit 1: Site Selection (1 Point)

The intention of this credit is to avoid development of inappropriate sites. In order to attain this credit, the building *cannot be built* on any of the following areas:

- Prime farmland (defined at http://www.farmland.org).
- Land that has never been developed *and* that is lower than 5 feet above the 100-year flood elevation (http://www.fema.gov).
- Land that is habitat for threatened or endangered species and identified as such on federal or state lists.
- Land that is within 100 feet of any wetlands, or state or local setbacks, whichever is more stringent.
- Land that has never been developed and is within 50 feet of a body of water.
- Land that was parkland prior to purchase.

The best way to gain this credit is to choose a previously developed site. Sweetwater Sound achieved it by building on a site located at 5501 U.S. Highway 30 West in Fort Wayne, Indiana, that had previously been developed by North American Van Lines, a national moving company. There was an existing 38,000-square-foot, one-story building that had been used as a sales and training facility. In addition to the building, the newly acquired Sweetwater site had a large paved parking area for semi-truck driver training and parking for home-based

Existing building in background, warehouse addition beginning in foreground. *Sweetwater Sound/John Hopkins. Reproduced by permission.*

trucks. The location of the site is near Interstate highway I-69, providing ideal access for truck traffic. The existing building has been reused and redeveloped without disturbing any previously undeveloped ground. Neighboring sites are all commercial and industrial facilities.

SS Credit 2: Development Density and Community Connectivity (1 Point)

The intention of this credit is to encourage development in urban areas that already have municipal service infrastructure available. This requirement can be fulfilled in one of two ways: by emphasizing the development density aspect of the credit or by emphasizing the community connectivity aspect. If we're emphasizing development density, we can build or renovate on a site that had already been developed *and* is located in an area with a density of 60,000 square foot per acre net. If the emphasis is on community connectivity, we must still be building or renovating a site that had already been developed, and the location must include the following features. It has to be within a half mile of a neighborhood that has a sufficiently dense population with a minimum of 10 units per acre net, *and* must be within a half mile of ten basic services (like hardware stores, banks, beauty salons, etc.) *with* pedestrian access.

Sweetwater Sound did not seek to achieve this credit. Although municipal service infrastructure is present, their neighboring sites are all commercial buildings and lack diversity of uses. They are located on a major highway on the near-northwest side of Fort Wayne, which provides easy access for the semi-truck deliveries that arrive and leave each day. They do provide many basic services for employees on-site, which cuts down on employees' needing to use their cars to run errands during their lunch break. One such service is a concierge, who will deal with errands for employees like dry-cleaning or laundry. There is also an on-site barber/hair salon, restaurant, and workout room. These additional services are highlighted in the chapter discussing Innovation and Design Credits.

SS Credit 3: Brownfield Redevelopment (1 Point)

The intention of this credit is to encourage development of sites that may have environmental contamination by prior occupants. This may or may not be the case, as sometimes the contamination is only a perceived possibility, rather than an actuality. The requirement is that either the site is documented as being contaminated or it is defined as such by a governmental agency (from local to federal). Use either an ASTM E1903-97 Phase II Environmental Site Assessment to document the contamination or a local cleanup program.

The Sweetwater Sound site was not designated as a brownfield.

SS Credit 4.1: Alternative Transportation: Public Transportation Access (1 Point)

The intention of this credit is to encourage the use of public transportation in order to reduce the environmental impact from cars. In order to achieve this, the site must be either within a half mile of a mass transit rail system or within a quarter mile of two bus stops.

Sweetwater Sound did not try for this credit. There is no mass transit rail system in Fort Wayne, and their location puts them outside of the municipal bus system routes. However, they have created innovative methods of reducing auto traffic for their employees, which we'll discuss in the section on Innovation and Design.

SS Credit 4.2: Alternative Transportation: Bicycle Storage and Changing Rooms (1 Point)

This credit repeats the intention to reduce the impact on the environment from automobile use. It requires that secure bike racks or storage is provided within 200 yards of the building entrance for at least 5 percent of all building users, *and* shower and changing facilities must be provided for 0.5 percent of full-time equivalent building users. For residential buildings, covered, secure bike storage facilities must be provided for at least 15 percent of occupants.

Sweetwater Sound achieved this credit. They installed a bike rack with 20 available spaces. Also see MSKTD drawing C4.1 at the end of this chapter.

SS Credit 4.3: Alternative Transportation: Low-Emitting and Fuel-Efficient Vehicles (1 Point)

Again, this credit repeats the intention to reduce the impact on the environment from automobile use. There are three ways this credit can be achieved. You can either choose to provide fuel-efficient and low-emitting, company-owned vehicles with preferred parking, or provide preferred parking for fuel-efficient vehicles, or install alternative fuel refilling stations. Check the American Council for an Energy Efficient Economy website at http://www.greenercars. org, for their annual vehicle-rating guide. To be considered for this credit, the car must achieve a minimum score of 40.

Sweetwater Sound achieved this credit by using the second option. They have provided 21 preferred parking spaces reserved for fuel-efficient vehicles. There are preferred spaces located near all three entrance doors. See the architectural drawing C2.1 at the end of the chapter.

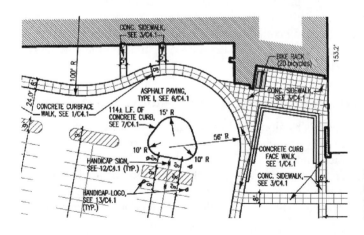

Drawings produced for Sweetwater Sound by MSKTD, an architectural firm headquartered in Fort Wayne, Indiana. Reproduced by permission.

Showers in changing room at Sweetwater Sound.   *Sweetwater Sound/John Hopkins.*
Reproduced by permission.

SS Credit 4.4: Alternative Transportation: Parking Capacity (1 Point)

Again, this credit has the intention of reducing the impact of environmental pollution by cars. It can be achieved by providing preferred parking for carpools and vanpools, and in addition, the total parking spaces for the building should be no more than local zoning indicates are necessary.

Sweetwater Sound achieved this credit. In addition to the preferred parking for fuel-efficient vehicles, they provide 21 spaces for car-pools and van-pools. These parking spaces are also divided amongst the three main entrance doors, with most of them being located near the employee entrance. Sweetwater also met the second part of this credit—having a total number of parking spaces that does not exceed the local zoning code. In order to do this, they had to remove a lot of pre-existing asphalt. The asphalt had been used by the previous owners as a training field for semi-truck drivers.

SS Credit 5.1: Site Development: Protect or Restore Habitat (1 Point)

The intention of this credit is to retain and restore as much natural green space as possible, so that habitat continues to be provided, or is restored for the creatures who share the land space with us. This intention is also to help sustain a biodiversity on the land. There are two possible methods of achieving this credit. The first is to limit compaction of greenfield sites. We cannot disturb the earth beyond 40 feet from the edge of the building, 10 feet beyond impervious

Much of the pre-existing asphalt was removed on the new Sweetwater Sound site.
Sweetwater Sound/John Hopkins. Reproduced by permission.

parking and sidewalk surfaces, and 25 feet beyond any pervious surfaces. The second option is for sites that were previously developed. With this option, at least 50 percent of the site must be restored or protected with plants that are native to the area.

Sweetwater Sound was not able to achieve this credit as the impervious area comprised approximately two-thirds of the building site. The developed land is approximately 14 acres on the corner of US Hwy 30 West and Kroemer Road. They own two adjacent parcels of land along Hwy 30, the middle one of which is approximately 15 acres of pre-existing asphalt parking lot, and the third parcel is approximately 15 acres of wooded land. The owners would eventually like to remove the asphalt from the middle parcel of land and restore it to natural green space; however, the cost of removing the asphalt is approximately one million dollars, and they'd prefer to defer that expense at this time. This is one of the credits they may decide to complete if their building comes within a point or two of being certified platinum.

SS Credit 5.2: Site Development: Maximize Open Space (1 Point)

The intention of this credit is to increase the ratio of open space to developed space beyond that which is mandated by local code. There are three ways to fulfill the requirements for this credit. The first is to exceed the open space required by local zoning code by 25 percent. In the case where there are no zoning laws that govern the site, the second way to achieve this credit is to keep open green space that is equal to the square footage of the building footprint. The third method is available for sites that have no local code requirements for open space retention. In this case, 20 percent of the site should be designated as green, open space.

Sweetwater Sound did achieve this credit. Our city code requires that a minimum of 10 percent of the building site be retained as open, green space. They were easily able to achieve 25 percent beyond code, as that would only have required them to retain 12.5 percent of green space. Sweetwater Sound was able to go well beyond this to achieve approximately 35 percent of open, green space on the building site. See the second MSKTD Drawing C2.1 at the end of this chapter for an illustration of the open, green space achieved.

SS Credit 6.1: Stormwater Design: Quantity Control (1 Point)

The intention of this credit in its essence is to keep the water on site that falls there, so the natural hydrology of the site is undisrupted. In order to achieve this credit, we have two cases to examine. The first case exists if the site has an area less than or equal to 50 percent that is impervious surface, and the second exists if there is greater than 50 percent impervious surface on the site. In the first case, we have to create and implement a plan that will ensure the post-development discharge rate is not greater than it had been before development, or that we protect any receiving streambeds by creating and implementing quantity controls. In the second case, we must develop a stormwater management plan that ensures a 25 percent reduction in runoff.

Sweetwater Sound did not achieve this credit. On the three parcels of land described earlier, there is a swale at the back of the property (on the south side), from which water moves in two directions. There was an existing small pond located at the southern juncture of the wooded and the asphalted acreage, to the west of the building site. Sweetwater Sound enlarged that pond, somewhat, to handle the water moving to the west. The water that moves to the east empties into a pond located on the land directly across Kroemer Road. One owner used to own both parcels of land—the land that Sweetwater Sound bought and the land parcel on the other side of Kroemer Road. There is a pre-existing agreement that allows water to move from the Sweetwater Sound land into the pond across the street. Despite the two ponds and the swale, the water handled does not equal a 25-percent reduction in runoff. The owners of Sweetwater Sound are still considering enlarging the pond on the southwest side of their property so that it is large enough to handle 25 percent of the water that falls on the developed land. The additional cost would be approximately 40,000 dollars. This is another credit they are likely to pursue if they are close to achieving LEED Platinum certification.

Swale at back of Sweetwater Sound property. Photograph taken by the Author.

SS Credit 6.2: Stormwater Design: Quality Control (1 Point)

The intention of this credit is to maintain the quality of the natural hydrology on the site by preventing pollution in the form of stormwater runoff from occurring. In order to attain this credit, a stormwater management plan must be implemented that captures at least 90 percent of the average yearly rainfall. The plan must identify how the stormwater will be held and treated and should utilize best management practices. Best management practices in treating stormwater runoff are those capable of removing 80 percent of total suspended solids following development. You can either refer to local or state specifications that utilize these standards or utilize equipment that monitors performance and demonstrates compliance with these criteria. The strategies that will work are measures that will reduce impervious surfaces on the site. You'll want to consider things like roof gardens, gravel grass for overflow parking, pervious pavement, rain gardens and/or bioswales, and constructed wetlands to satisfy this credit.

Sweetwater Sound did not achieve this credit. Too much water falls on this land to be handled through a natural system. A stormwater filtration system will have to be installed in the swale on the south side of the property. It will be a swirl tank that will remove all pollutants before the water flows in to the ponds. The additional cost will be approximately 80,000 dollars, and is another item being considered if they are close to achieving LEED platinum certification.

SS Credit 7.1: Heat Island Effect: Non-Roof (1 Point)

The intention of this credit is to reduce the effect of heat islands, which are caused by dark surfaces on roofs, pavement, and parking lots. These dark surfaces absorb the heat from the sun and elevate the temperature in the surrounding area. This can cause areas that have a high incidence of black roofs and parking lots, typical of cities and towns, to be 10–12° Fahrenheit higher on summer days. This affects us negatively by increasing the demand for electricity during peak hours, increases the cost of air conditioning, adversely affects the environment by increasing air pollution, and causes occurrences of heat-related illnesses and deaths to increase.[2] This is also a global warming issue. As we increase areas of the earth's temperature, we are melting the polar ice pack.

There are two possible ways to attain this credit. One is to supply at least 50 percent of the hard surfaces with trees sized to provide shade within five years, and/or utilize paving materials that have a high solar reflectance index (at least 29), and/or utilize an open grid pavement system.

Sweetwater Sound did not achieve this credit. They investigated a grass paver system for the parking areas, but discovered that the material underneath of the existing parking lot had no stone—it was solid clay. They would have had to remove the clay and put in stone with an estimated cost of several million dollars. They also investigated replacing the existing asphalt with highly reflective concrete, but this too would have cost in the range of four million dollars. They also investigated planting trees in the parking lot, but this would have been too expensive as well, not because of the trees themselves, but because our local code would require the installation of islands with concrete curbs.

SS Credit 7.2: Heat Island Effect: Roof (1 Point)

The intention of this credit is the same as the above. It is meant to reduce the effect of heat islands. There are three possible ways to achieve this credit. The first is by using roofing materials, on at least 75 percent of the roof surface, that supply a high solar reflective index: a minimum of 78 for roofs with a slope of less than or equal to 2:12 (low sloped), and a minimum of 29 for roofs with a slope of greater than 2:12. The second method is covering at least 50 percent of the roof area with a green, vegetated roof. The third method is to install a combination of the two.

Sweetwater Sound did achieve this credit. One hundred percent of the roof surface is covered with light colored, high albedo roofing membrane having a solar reflective index of 111. The roofing system is Carlisle's Sure-Flex Polyvinyl Chloride (PVC) membrane. The owners understand the problem with using PVC, and investigated using Carlisle's Sure-Weld® thermoplastic polyolefin (TPO) membrane. At the time, the TPO roofing system was so new that it did not come with any warranties, while the PVC system carried a 25-year warranty. For that reason, they decided to go with the PVC system. On the positive side, they did replace the black rubber Ethylene Propylene Diene Monomer (EPDM) roof that was on the existing building with a new high-albedo white roof.

Sweetwater Sound/John Hopkins. Reproduced by permission.

SS Credit 8: Light Pollution Reduction (1 Point)

The intention of this credit is to maintain or, in many cases, to begin to recover the visibility of the night skyscape. It also intends to protect land surrounding the site from light trespass for inhabitants of those areas. In order to achieve this credit, attention must be paid to both interior and exterior lighting. There are two ways of addressing interior lighting. Either the interior lighting must be on an automatic control that turns the lights off during hours when the building is not used, with a manual override capacity, or the maximum candela of the interior lights does not exit through windows. For exterior lighting, the Exterior Lighting Section of ASHRAE/ IESNA Standard 90.1-2004 should be followed. The standard specifies that exterior area lighting densities should not exceed 80 percent, and building facades and other features should not exceed 50 percent. Furthermore, the exterior lighting should follow the requirements of IESNA RP-33 defined zone within which the project is located. IESNA RP-33 provides standards for outdoor environmental lighting and is meant to address issues of light pollution. This standard identifies zones as "dark" for rural and park settings, "low" for residential areas, "medium" for commercial and industrial settings, and "high" for city centers. For more specific information, this standard can be purchased online.

Sweetwater Sound did achieve this credit. Full cutoff luminaires were used for all exterior lighting. This is something that the county building code already requires. They also have less than 50 percent of building features lighted after dark.

Sweetwater Sound—front entrance after dark.   *Sweetwater Sound/John Hopkins. Reproduced by permission.*

Summary

This chapter describes the conventional approach to building regarding the way the site itself is treated, and explains the different approach taken if we consider the site with sustainability in mind. In green building, we consider the total ecological context in which the site exists first, and find ways for the ecological conditions to retain and sustain the functions that existed prior to human building. One of the reasons we do this is because these ecological conditions are the fundamental framework needed for human life to be sustained.

The chapter gives the steps that need to be taken to collect data about the site, and includes site research components of conducting wetland, floodplain, and deed research, as well as capturing existing aerial photos. Directions are also given for the kind of data needed that can only be acquired by conducting an actual site visit, and includes actions like identifying and shooting a topographical survey of existing trees, and talking to neighbors to gain knowledge about the site that may not be apparent under the conditions of your site visit. Directions are also given for the third aspect of data collection, which is energy mapping and modeling.

This chapter introduces a commercial project that we will follow through its construction and submittal process as it seeks LEED NC Gold certification. The chapter explains LEED Sustainable Sites by following the project through each prerequisite and credit, noting what they were able to achieve, what they couldn't achieve, and the reasons why or why not. An addendum at the end of the chapter is the LEED scorecard and cost analysis the LEED-AP and owners used during this project.

Questions and/or Assignments

1. What is a biome?
2. What is an ecosystem?

3. Is it possible to build in such a way that the building could behave in the same way, regarding carbon dioxide and oxygen, as a conifer forest would? How?
4. What are some methods we could use to keep the water on the land that falls there?
5. Conduct the site research on a specific property: include wetland, floodplain, and deed research.
6. Conduct energy mapping and modeling on that same property.
7. Using the property from questions 1 and 2, take it through each of the Sustainable Sites prerequisites and credits and assess the properties ability to satisfy them.
8. Do the prerequisites have to be satisfied in each area in order to achieve certification?
9. What must the Erosion and Sedimentation Control plan comply with?
10. What is a brownfield?
11. Which standard specifies the exterior lighting densities?

Notes

1. Mazria, Edward. *The Passive Solar Energy Book*. Emmaus, PA: Rodale Press, 1979.
2. See http://www.epa.gov/hiri/.

Sweetwater LEED scorecard / cost analysis

11/5/2007

Sustainable Site

	7	0	0	7							
	Y	Y?	N?	N				initial cost	annual return	payback	notes
	Y				Prereq		Construction Activity Pollution Prevention	n/c			Already included due to IDEM requirements
*	1				Credit	1	Site Selection	n/c			Site was previously developed
	1				Credit	2	Development Density & Community Connectivity	n/c			
	1				Credit	3	Brownfield Redevelopment	n/a			Not a hazardous site
	1				Credit	4.1	Public Transportation Access	n/a			Buses not available (not per Citilink maps)
	1				Credit	4.2	Bicycle Storage & Changing Rooms	600	n/a		Need racks for 5% of building users near an entry (showers covered in fitness area)
	1				Credit	4.3	Low-Emitting & Fuel Efficient Vehicles	625	n/a		Preferred parking for 5% of total parking
	1				Credit	4.4	Parking Capacity	n/c			Parking based on use (1/300sf office, 1/2000 warehouse . . .)
											Provide prefered parking for car poolers (5% of total count)
	1				Credit	5.1	Site Development, Protect or Restore Habitat	n/a			Must restore 50% of non- building area to natural vegetation
	1				Credit	5.2	Site Development, Maximize Open Space	n/c			10% of parking area required to be landscape area. Must be area inside r.o.w. lines
											Include only currently developed project area
	1				Credit	6.1	Stormwater Design, Quantity Control				Reduce stormwater runoff by 25% (re-use rainwater . . . very costly @ $100k, use pourous pavement . . . costly and doesn't drain in clay soils)
	1				Credit	6.2	Stormwater Design, Quality Control	n/a			Filter on storm line.
	1				Credit	7.1	Heat Island Effect, Non-Roof	n/a			Shade 50% of all hardscape, use concrete pavement
	1				Credit	7.2	Heat Islands Effect, Roof	30,600	$$$		PVC (or metal) with reflectance of SRI 79 instead of epdm on 75% of roof
	1				Credit	8	Light Pollution Reduction	n/c			Cutoff lighting required by City already (LZ 3 requirements)
											Less than 50% of exterior bldg./features lighted

Y	Y?	N?	N		No.					
5	0	0	0	**Water Efficiency**						
1				Credit	1.1	Efficient Landscaping, Reduce by 50%				High efficiency sprinklers, reduce area, select native or adapted plants
1				Credit	1.2	Efficient Landscaping, No Potable Use and No Irrigation				No permanet sprinklers (remove after one year)
1				Credit	2	Innovative Wastewater Technologies				Reduce potable water use for sewage by 50%
										Review low flow shower heads
1				Credit	3.1	Water Use Reduction, 20% Reduction	1,250	2,054	0.61	Current calculation is 54% reduction (kitchen need not be considered)
										Waterless urinals, dual flush waterclosets ($40 ea more) @ women's
1				Credit	3.2	Water Use Reduction, 30% Reduction	in 3.1			Low water usage faucets (hands free)
8	4	2	3	**Energy and Atmospere**						
1				Prereq	1	Fundamental Commissioning	96,250	20,000	4.81	Checks design and installation (balancing, tc....) to maximize performance (5 to 10% more efficient)
1				Prereq	2	Minimum Energy Performance		n/c		Meet ASHRAE 90.1 (baseline ofr 1.1) (have confirmed cooling tower, boilers and chillers)
1				Prereq	3	Fundamental Refrigerant Management		n/c		Non CFC refrigerants required by code
1				Credit	1.1	Optimize Energy Performance 10.5%				Need to create base model
1				Credit	1.1a	Optimize Energy Performance 14%				Current performance estimate is 17% excluding daylight harvesting savings
1				Credit	1.2	Optimize Energy Performance 17.5%				
1				Credit	1.2a	Optimize Energy Performance 21%				
1				Credit	1.3	Optimize Energy Performance 24.5%				
	1			Credit	1.3a	Optimize Energy Performance 28%				
	1			Credit	1.4	Optimize Energy Performance 31.5%				
	1			Credit	1.4a	Optimize Energy Performance 35%				
	1			Credit	1.5	Optimize Energy Performance 38.5%				
		1		Credit	1.5a	Optimize Energy Performance 42%,				

(continued)

(continued)

Y	Y?	N?	N	pts	Type	No.	Item	Cost	Value	Value	Value	Notes
							Upgraded glass		119,200	44,400	2.68	
							Skylights		120,000			
							HVAC tied into lighting systems	$$$				
							High efficiency boilers		40,000	10,000	4.00	
							Carbon dioxide sensors		2,500			
							Ceiling fans in warehouse	$$$				
							Dimming systems	$$$				
				1			Motion detectors	$$$				
					Credit	2.1	On-Site Renewable energy, 2.5%					Windmills . . . . 50kw (3%), $150,000, annual return 6k, possible grant
				1	Credit	2.2	On-Site Renewable energy, 7.5%					Windmills . . . . 3x50kw (9%), $450,000, annual return 18k, possible grant
				1	Credit	2.3	On-Site Renewable energy, 12.5%	n/a				
	1			1	Credit	3	Enhanced Commissioning					Additional involvement esp. with owner training and follow through . . . $23,000
				1	Credit	4	Enhanced Refrigerant Management	n/a				Ice storage system requires HCFC R123 usage
1					Credit	5	Measurement & Verification	n/c				Monitor energy use to increase long term efficiencies
1				1	Credit	6	Green Power		3,600	n/a		2year contract, provide 35%
8	0	0		5	**Materials & Resources**							
Y	Y?	N?	N									
Y				1	Prereq	1	Storage & Collection of Recyclables	n/c				
				1	Credit	1.1	Building Reuse, 75% of Existing Walls, Roofs, Floors	n/a				Max. new bldg/existing is exceeded
				1	Credit	1.2	Building Reuse, 95% of Existing Walls, Roofs, Floors	n/a				
				1	Credit	1.3	Building Reuse, 50% of Interior Non-Structural Elements	n/a				
					Credit	2.1	Construction Waste Management, Divert 50%	$$$				
				1	Credit	2.2	Construction Waste Management, Divert 75%	$$$				
					Credit	3.1	Materials Reuse, Specify 5%	n/a				
				1	Credit	3.2	Materials Reuse, Specify 10%	n/a				

	Y?	N?	N						
1				Credit	4.1	Recycled Content, Specify 10%	n/c		
1				Credit	4.2	Recycled Content, Specify 20%	n/c		
1				Credit	5.1	Regional Materials, 10% Manufactured Regionally	n/c		
1				Credit	5.2	Regional Materials, 20% Extracted Regionally	n/c		
1				Credit	6	Rapidly Renewable Materials	$$$		2.5% of building materials ($22,500)
1				Credit	7	Certified Wood	2,500	n/a	Min. 50% FSC of wood based materials (add 2% to cost of regular wood)

Indoor Environmental Quality

11	2	1	1	1					
Y	Y?	N?	N						
Y				Prereq	1	Minimum IAQ Performance	n/c		Meet ASHRAE 62.1
Y				Prereq	2	Environmental Tobacco Smoke (ETS) Control	n/c		
1				Credit	1	Outdoor Air Delivery Monitoring	$$$		Carbon dioxide sensors
	1			Credit	2	Increase Ventilation	n/a		Increase ventilation by 30% over ASHRAE 62.1
1				Credit	3.1	Construction IAQ Mgmt Plan, During Construction	$$$		
1				Credit	3.2	Construction IAQ Mgmt Plan, Before Occupancy	$$$		
1				Credit	4.1	Low-Emitting Materials, Adhesives & Sealants	$$$		
	1			Credit	4.2	Low-Emitting Materials, Paints & Coatings	$$$		Update on paint VOC situation
1				Credit	4.3	Low-Emitting Materials, Carpet Systems	$$$		
1				Credit	4.4	Low-Emitting Materials, Composite Wood & Agrifiber Products	$$$		
1				Credit	5	Indoor Chemical & Pollutant Source Control	$$$		Walkoff mats min. 6', filters
1				Credit	6.1	Controllability of Systems, Lighting		10,000	Add task lights at work stations
		1		Credit	6.2	Controllability of Systems, Thermal Comfort	n/a		Clarifying amount of control needed . . . at 50% of cubicles?
1				Credit	7.1	Thermal Comfort, Design	n/c		Meet ASHRAE 55
1				Credit	7.2	Thermal Comfort, Verification	$$$		Survey of employees, correct if 20% dissatisfied

(continued)

(continued)

Y	Y?	N?	N								
1				Credit	8.1	Daylight, 75% of Spaces	in EA				
1				Credit	8.2	Views, 90% of Spaces	n/c				Direct line of sight in 90%
5	**0**	**0**	**0**			**Innovation & Design Process**					
Y	Y?	N?	N								
1				Credit	1.1	Innovation in Design: Ice storage / reduce peak elec. usage	20,500		6,000	3.42	
1				Credit	1.2	Innovation in Design: Educational Building	-				Some education materials and tour time
1				Credit	1.3	Innovation in Design: Water use reduction 40%	in 3.1				
1				Credit	1.4	Innovation in Design: Heat island effect roof	in SS7.2				100% of roof
1 (*)				Credit	2	LEED Accredited Professional	n/c				
						Innovation in Design: Green housekeeping methods					
						Innovation in Design: In-house dining (reduces travel)					
						Innovation in Design: 15% open space (43% actual)					
44	**6**	**3**	**16**			Goal of 39 for gold (platinum 52)					
						* items have documention submitted					
						Documentation costs	40,000				
						Energy modeling beyond standard services			26,000		
						Total costs	513,625		82,454	6.23	

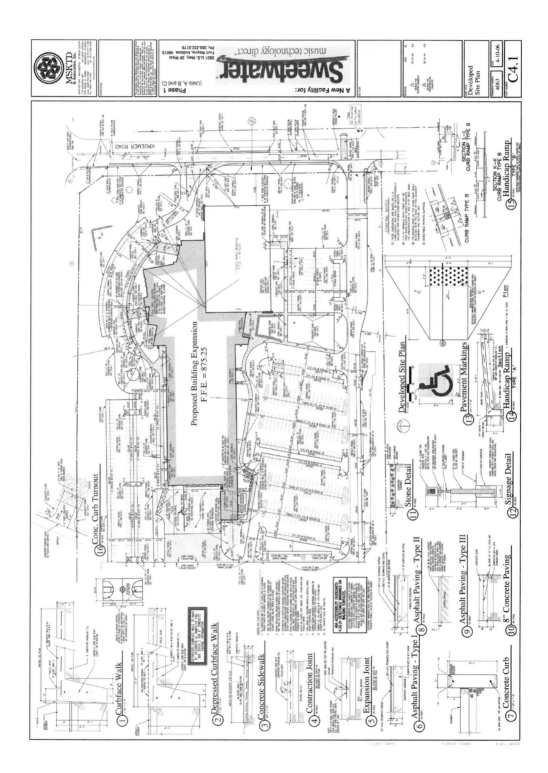

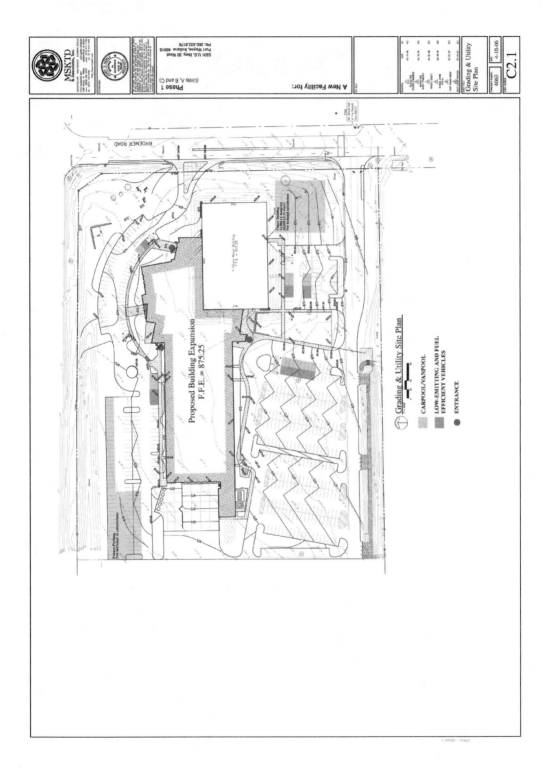

Grading & Utility Site Plan

- CARPOOL/VANPOOL
- LOW-EMITTING AND FUEL EFFICIENT VEHICLES
- ● ENTRANCE

Proposed Building Expansion
F.F.E. = 875.25

KROEMER ROAD

MSKTD
& Associates, Inc.

5501 U.S. Hwy. 30 West
Fort Wayne, Indiana 46818
Ph: 260.432.8176

A New Facility for:

Phase 1
(Units A, B and C)

Grading & Utility
Site Plan

C2.1

4063   4-10-06

68

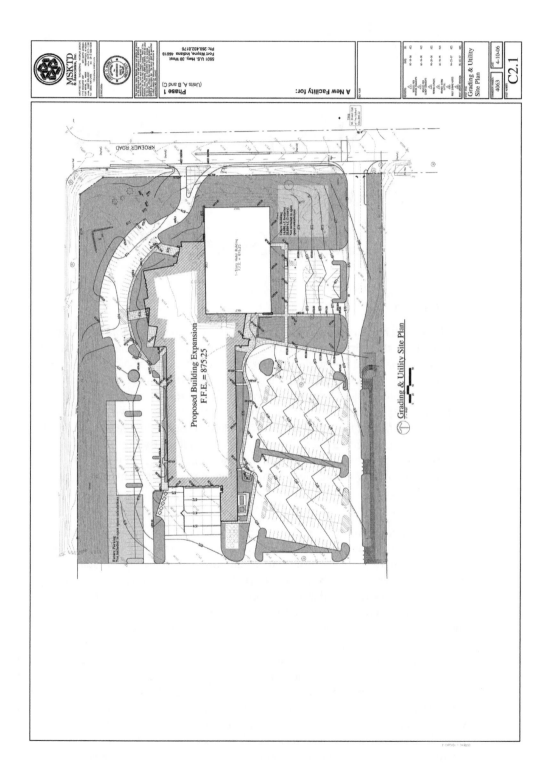

Proposed Building Expansion
F.F.E. = 875.25

KROEMER ROAD

Grading & Utility Site Plan

MSKTD
& Associates, Inc.

5501 U.S. Hwy. 30 West
Fort Wayne, Indiana 46818
Ph: 260.432.8176

A New Facility for:

Phase 1
(Units A, B and C)

Grading & Utility
Site Plan

C2.1

4-10-06
40063

7

■ ■ ■

Water Resources and Sustainable Landscaping

Abstract

This chapter is concerned with water resources and sustainable landscaping. It explains the minimum amount of water that human beings must have in order to maintain life, and explains what we must do in order to maintain our Earth's potable water resources. We must build in such a way that the Earth's water tables are recharged rather than diminished by the actions we take. The chapter looks at three types of water flows: the water that flows into the building, the water that flows out of the building as effluent, and the water that falls onto the site. The chapter also follows the commercial project introduced in the last chapter through the LEED credits in water efficiency, noting what they were able to achieve together with an explanation of how they did it.

Most of us understand now that water is a precious, but diminishing resource that we must think about and value with respect. The news is filled each day with reports of different areas on the Earth that suffer from lack of potable water. We read about lakes and aquifers drying up at an astounding rate because of the way in which we have used and are using potable water. Even areas of the United States (Atlanta, GA) are now suffering from shortages of potable water. In some areas of the southwestern United States, water is being taken from aquifers at twice the rate of natural recharge. One way of beginning to look for solutions to this problem is to know how much water one human being must have in order to maintain life each day, so that we can begin to understand our planetary needs. Human beings must have between 0.5 and 1 gallon of water each day for drinking and 1 gallon each day for food preparation. In many areas of the world, people exist on just this much water. In the United States, each person consumes an average of 1,800 gallons of potable water each day. Much of the potable water that is calculated into this average is consumed by buildings in unnecessary and wasteful ways.

There is a lot that we can do to change that. There are three water flows that we should investigate in order to create more efficiency: water that flows into the building, water that flows out of the building, and water that falls onto the site. In the conventional method of building, potable water is used for everything from flushing toilets to watering lawns. Furthermore, all the rain water that falls on the land is moved into and through the storm sewers as quickly as possible. From the storm sewer, the water is moved into creeks, rivers, and lakes, and much of it eventually ends up in the ocean. Our task in sustainable building is to design ways to capture, store, and use rainwater on-site. We can use simple techniques involving rain barrels and rain gardens, or more complex systems using cisterns and bioswales. If we are able to capture large amounts of rainwater, we can filter it and reuse it in the building for tasks like flushing toilets and watering landscaping.

The water that exits as effluent from our homes and buildings typically moves through a different sewer line to a centrally located treatment facility that is often miles from where we live and work. We can design ways to treat the water that exits our homes and buildings with filtering systems capable of cleaning the water and allowing it to move back to the water table. These systems can be as simple as the old-fashioned septic systems, or as complex as living machines that can handle large amounts of effluent.

USGBC LEED-NC WATER EFFICIENCY, FIVE POSSIBLE POINTS

WE Credit 1.1: Water Efficient Landscaping: Reduce by 50 Percent (1 Point)

The intention of this credit is to create landscaping in such a way that it does not need augmentation with water to survive and thrive. This credit calls for a 50 percent reduction of potable water utilized for irrigation purposes, measuring against a baseline case of mid-summer usage. This credit can be achieved in any combination of a number of ways. One strategy would be to plant species that are native to the area. Native species develop root systems that help the plants to thrive in the soil and weather conditions natural to the area. Other helpful strategies would be to utilize methods of irrigation that are efficient: capturing rainwater in cisterns or rain barrels to use for irrigation, and using gray water effluent from the building for irrigation.

Sweetwater Sound did achieve this credit. They planted native plant species, and are using high-efficiency sprinklers for the first year to help the plant roots get well-seated.

WE Credit 1.2: Water Efficient Landscaping: No Potable Water Use or No Irrigation (1 Point in Addition to WE Credit 1.1)

The intention of this credit is to entirely remove the need to use potable or other water resources for irrigation purposes. This credit can be attained by making sure that no potable water is used at all for landscape irrigation. Only rainwater or recycled water may be used for landscape needs. And while an irrigation

Native Landscaping Being Installed at the Sweetwater Sound Entrance.   Photograph by Author.

system may be installed to assist plantings in taking root, the system must be removed within one year.

Sweetwater Sound did achieve this credit, as their irrigation system is being removed within one year of installation.

WE Credit 2: Innovative Wastewater Technologies (1 Point)

The intention of this credit is to create ways to recharge the local aquifer, and at the same time, reduce the use of potable water and reduce the amount of wastewater that gets generated. There are two ways to achieve this credit. One way is to reduce the use of potable water by 50 percent for flushing toilets or urinals, and the other way is to treat and use wastewater on-site. To achieve this in the first method mentioned above, either we can use waterless urinals and water-conserving toilets to reduce potable water use or we can create rainwater or gray water cachement systems and plumb the nonpotable water to water closets. In order to achieve this credit using the second method, a created method of treating at least 50 percent of the wastewater on-site would be necessary. This can be done by constructing a filtering system and wetlands that would remove the biological nutrients and allow the cleaned water to make its way back to the water table.

Sweetwater Sound achieved this credit by installing waterless urinals and efficient toilets that can be flushed in two ways—one that takes a little more water, and the other, a little less. Normally, to reduce the wastewater enough to satisfy the requirements of this credit, it requires the construction of at least some gray water cachement system. However, Sweetwater Sound has an employee population that is 80 percent male, so using the combination of waterless urinals and ultralow-flow fixtures meant they were able to attain this credit without having to install a gray water system. They achieved a 54 percent reduction in potable water use, with an increase in first costs of just 40 dollars per water closet in the women's restrooms. The initial cost was 1,250 dollars, and the annual return in terms of savings on operating costs of water is 2,054 dollars. Thus, the initial expense is being recovered inside of the first year of operation.

WE Credit 3.1: Water Use Reduction: 20 Percent Reduction (1 Point); WE Credit 3.2: Water Use Reduction: 30 Percent Reduction (1 Point)

The intention of these two credits is to create additional ways to relieve city infrastructure wastewater and water supply systems. In order to achieve one or both of these credits (using fixtures that satisfy the Energy Policy Act of 1992 for performance—the credits are calculated using these fixtures as the baseline for the building's water use), use strategies regarding toilets, urinals, showers, and bathroom and kitchen sink faucets to create a 20–30 percent or more reduction in potable water consumption. Strategies might include using high-efficiency

Waterless Urinals in the Men's Restroom at Sweetwater Sound.   *Sweetwater Sound/John Hopkins Reproduced by Permission.*

Dual-Flush Water Closets at Sweetwater Sound. Photographs Taken by the Author.

fixtures that surpass the Energy Policy Act of 1992, and/or plumbing the building with a gray water system that would use captured rain water and/or recycled and filtered water from bathroom sinks and showers to flush toilets. Where code allows, waterless urinals and composting toilets are good alternatives, as well.

Sweetwater Sound achieved both of these credits by installing the waterless urinals, dual flush water closets, and hands-free, ultralow-flow faucets with infrared sensors in bathrooms. The ultralow-flow turbine inside of the sink recharges the battery that fuels the infrared sensors.

Ultralow-Flow Faucet with Infrared Sensor at Sweetwater Sound. Photograph Taken by the Author.

Summary

This chapter is concerned with water resources and sustainable landscaping. It explains that human beings must have a minimum of one-half to one gallon of water each day to drink and one gallon of water each day for food preparation in order to maintain a human life, and that the potable water resources of the Earth are diminishing. In order to maintain and even increase potable water reserves, we must build in such a way that the Earth's water tables are recharged rather than diminished by the actions we take. The chapter explains that we must treat three types of water flows as resources: the water that flows into the building, the water that flows out of the building as effluent, and the water that falls onto the site.

This chapter also follows the commercial project introduced in the previous chapter through the LEED credits in water efficiency, noting what they were able to achieve together with an explanation of how they did it. Our project was able to achieve all five of the possible water credits.

Questions and/or Assignments

1. Do some research and find out about the water table in your area. Write a brief paragraph about what you discover, and cite your references.
2. Do some research to discover the source and quality of the potable water that comes through your faucet. Does it come from a reservoir, river, or water table? Also investigate how the source of your drinking water gets replenished.
3. Find out what the native plant species are in your area. Discover what qualities the plants have that are conducive to their flourishing in the specific climate and Earth crust conditions in your area.
4. Design a plan for "your yard" (no matter where you live) that will capture and utilize rainwater.
5. Design a plan of plantings for "your yard" that will require no watering after the first year.

8

■ ■ ■

Building Orientation, Renewable Energy and Storage, and HVAC Systems

Abstract

This chapter is concerned with orienting the building, capturing and storing renewable energy, and HVAC systems. It discusses the problems encountered, especially with humidity, in trying to design larger buildings with the capacity for passive cooling. We follow our commercial project through the LEED prerequisites and credits in energy and atmosphere, noting what they were able to achieve and what they weren't, with an explanation of why or why not.

In the conventional way of building, thinking about orientation in order to capture natural energies doesn't usually enter into the picture. The buildings are oriented to face the street, and the driveways are designed to be as short and straight as possible so that vehicles can be accommodated. Our task in sustainable design and construction is to capture the sun so that it can be used for passive solar heating in the winter, while at the same time the structure guards against cold winds. We also want to capture desirable breezes and create shade to prevent solar gain in warm weather. Major roof areas should be designed to face the direction that will be able to harvest the most energy from the sun as possible, for installation of solar hot water heater and/or photovoltaic panels.

With a large commercial building, it is more important to think about methods of passively cooling the structure. But there are many difficulties with this approach. In the case of Sweetwater Sound, no mechanical engineers in our region would consider natural ventilation on a building of this size because the humidity is too hard to regulate and manage using natural systems. They don't like the idea that people could open windows at inappropriate times, make the mechanical systems expend more energy, and throw the humidity levels off balance. This was frustrating to one of the architects who worked on the Sweetwater Sound project. He envisioned the possibility of an automated system that would allow windows to be opened when the outside temperature and humidity levels are appropriate, and not otherwise, but none of the mechanical engineers would consider it.

This is easier to think about and create natural, passive systems for residential applications and smaller buildings. The Adam Joseph Lewis Center for Environmental Studies at Oberlin College houses classrooms, faculty offices, and a living machine that process all of the wastewater generated by the building. It uses both active mechanical and passive systems to ventilate the building. There are 24 geothermal loops that exchange the heat with the ground before sending it on to the heat pump. Energy recovery ventilators (ERVs) draw fresh air into the building. As the ERV moves past the air that is exiting the building, the heat between the ERV and air is exchanged so the air is pre-tempered before it is sent on to the heat pump to be distributed. Convection cooling also occurs in this building. In the Atrium, pictured below, there are window vents along the top row of windows which automatically open when conditions are desirable.

The conventional method of building uses power from the gas and electric grid, and doesn't normally consider anything else, with the possible exception of geothermal heating and cooling. The vast majority of HVAC

The Atrium of the AJL Center for Environmental Education at Oberlin College. Reproduced by Permission.

systems installed in homes and buildings in the United States are forced air furnaces and air conditioners. These forced air systems heat and cool air, which makes them tremendously inefficient methods of conditioning the air in a building. Air lacks the density necessary to hold heating or cooling and must be continuously re-heated or re-cooled in order to maintain comfort levels. In sustainable construction, we build into the design passive forms of heating, cooling, and daylighting, and augment these with efficient systems that heat and cool mass rather than air.

USGBC LEED-NC ENERGY AND ATMOSPHERE, 3 PREREQUISITES AND 17 POSSIBLE POINTS

EA Prerequisite 1: Fundamental Commissioning of the Building Energy Systems (Required)

The intention of this prerequisite is to make certain that all HVAC systems function as they are meant to do. The commissioning agent must see to it that these systems are installed, calibrated, and perform according to the owner's specifications. Commissioning is one of the most important things that can be done to make certain that a building will perform at its absolute best in terms of energy efficiency.

In order to achieve this prerequisite, the commissioning team must perform the following tasks. First: A qualified commissioning agent (CxA) should be chosen. This person must have experience as a CxA on no fewer than two prior construction projects, must be responsible solely to the owner to report all findings, must be independent of the project management and design teams (even though this person may be employed by the same firm), and must be responsible for all commissioning activities from beginning to end. One caveat to these requirements is that if the project is smaller than 50,000 square feet, a qualified CxA may be chosen from the design or management team, but they are still responsible solely to the owner. Second: The CxA is responsible to review the owner's project requirements (OPRs) and the basis of design (BOD) documents to ensure they are both complete and clear. (The owner and design team are responsible for any updates to these documents.) Third: The CxA must write the commissioning requirements and make sure they are incorporated into the construction documents. And finally: The CxA is responsible for developing and implementing a commissioning plan, to verify that the systems perform correctly, and to write a commissioning report for the owner.

The systems that must be commissioned are all energy systems along with their controls, whether passive or mechanical, that heat, ventilate, air-condition, or refrigerate the building, all controls for light providing systems, all hot-water systems, and all systems for renewable energy supply.

Sweetwater Sound satisfied all energy and atmosphere prerequisites. A CxA was hired to commission the building energy systems. Doing so cost one dollar per square foot, but the improvement in energy efficiency achieved was between 5 and 10 percent on the various systems. Their initial investment is estimated to be returned to them in just under five years of operation.

EA Prerequisite 2: Minimum Energy Performance (Required)

The intention of this prerequisite is to set the minimum level of energy efficiency that is acceptable for the building's systems. Essentially, all projects must comply with (or if the project has been/is being registered after June 26, 2007, must exceed) ASHRAE/IESNA Standard 90.1-2004. The strategy for achieving this prerequisite is to design the building envelope and all energy-related systems to achieve the highest possible energy performance. If you refer to the User's Manual for ASHRAE 90.1-2004, you will find worksheets to assist in documentation for this prerequisite.

Sweetwater Sound achieved this prerequisite by meeting ASHRAE 90.1-2004. They optimized energy performance in the building envelope, HVAC systems (boilers, chillers, and air handlers), and by harvesting site energy in the form of daylighting.

EA Prerequisite 3: Fundamental Refrigerant Management (Required)

The intention of this prerequisite is simply and importantly to reduce ozone depletion. In order to do that, it is required that no chlorofluorocarbon (CFC)-based refrigerants be used in any new HVAC&R systems. Furthermore, if the new construction is being added to an existing building that uses a CFC-based system, it must be phased out and converted before the project is completed.

Sweetwater Sound achieved this prerequisite by using non-CFC refrigerants.

Calmac Ice Storage system.   *Sweetwater Sound/John Hopkins. Reproduced by Permission.*

EA Credit 1: Optimize Energy Performance 1–10 Points (Two Points are Required for LEED-NC Projects Registered after June 26, 2007)

The intention of this credit is to improve energy efficiency beyond the baseline established in EA Prerequisite 2. There are three possible ways to acquire points in this credit. One way is to conduct whole building energy simulation, and the other two have to do with following two different prescriptive compliance paths. With the whole building energy simulation method for new buildings, the required two points are gained by achieving 14 percent building performance improvement over baseline, which augments in increments of 3.5 percent per additional credit point, up to the maximum of ten points for achieving 42 percent building performance improvement over baseline. In this energy analysis, all energy costs must be taken into account, using the default baseline building process energy cost of 25 percent. (Process energy loads include items like computers, elevators, kitchen and laundry equipment, and the alike.)

The first of the prescriptive options follows the ASHRAE Advanced Energy Design Guide for Small Office Buildings 2004, and applies only to buildings that are fewer than 20,000 square feet, designated as office space only, and the criteria that must be followed are climate zone specific. A maximum of four points are possible to achieve using this method. The second of the prescriptive options requires that the Advanced Buildings Benchmark Version 1.1 be followed. But with this option, only one point is achievable, so at this point it doesn't comply with the required two points necessary. It is currently being worked on and updates to this can be found on the Credit Interpretation (CIR) page of the USGBC website.

Sweetwater Sound used the energy simulation model to attain these points. They have definitely achieved between eight and ten of the points available by optimizing energy performance by 35–42 percent over the baseline model. The software they used is "eQUEST" and is freeware available through the U.S. Department of Energy website: http://www.doe2.com/. The software is a tool designed specifically to help you analyze and compare building designs and their mechanical systems as to their energy use and efficiency. The site claims that no special experience in building performance modeling is necessary, but it does take time to enter the data. For a building the size of Sweetwater Sound, it took one person a full week to enter the data and build the building as a model in the eQUEST software. First, build the model in the software using the minimum ASHRAE requirements to establish the baseline, and then enter the features of the building being constructed that are different from the baseline.

All building surface R-values must be entered: foundation, roof, slab, and walls. The Sweetwater Sound roof achieved an R-value of 30, stone and stud walls, an R-value of 25, warehouse solid surface walls, an R-value of 33, and the translucent wall panels (used to help achieve daylighting in the warehouse) have an R-value of 7. The existing building roof was improved from the original R-value of 15 to an R-value of 30, and the walls went from an R-7 to an R-30 (except the fenestration). The windows throughout the building are 25 percent better than required by the ASHRAE minimum.

All mechanical equipment, models and efficiency levels, also must be entered into the eQUEST software. For Sweetwater Sound, that meant entering information on the boilers, chillers, air handlers, temperature set points and hours of operation, the building's electrical consumption, and so on. Sweetwater Sound has a variable air volume system supported by a central chilled water plant, boilers, 15 zones, and 15 air handlers that push air into the various zones.

According to the software data, Sweetwater Sound should achieve all points allocated for optimized energy performance. The actions that Sweetwater Sound has taken above and beyond what is required by ASHRAE that enabled them to achieve these points are as follows. They used more energy-efficient glass for an extra cost of 119,200 dollars with an expected return on investment of just over two and one-half years. The glass has a low-emissivity coating that reduces solar heat gain and glare, which reduces the cooling load and increases the comfort of building occupants. The skylights installed in the warehouse and in the large sales area are more energy efficient as well, and the upgrade cost was 120,000 dollars.

The HVAC systems are tied to the lighting systems, making a difference in the energy efficiency of the building. Highly efficient boilers added an initial cost of 40,000 dollars, saves 10,000 dollars per year in operating expense, and has an expected return on investment of just four years. Carbon dioxide sensors added 2,500 dollars to the initial cost of the building. Ceiling fans were installed in the warehouse, which creates energy efficiencies. Light sensors, which adjust

One of a Number of Skylights that Provide Daylight at Sweetwater Sound. *Sweetwater Sound/John Hopkins. Reproduced by Permission.*

Mechanical Room at Sweetwater Sound.   *Sweetwater Sound/John Hopkins. Reproduced by Permission.*

automatically to natural light levels and occupancy, and motion detectors are installed throughout the building, both of which add significantly to the energy efficiencies attained.

EA Credit 2: On-Site Renewable Energy (1–3 Points)

The intention of this credit is to reduce our automatic reliance on fossil fuel to provide energy for buildings, and to increase the use of renewable energy. In order to gain some or all of these credits, enough renewable energy must be used to supply from 2.5 percent to 12.5 percent of the buildings' projected yearly energy use. You can either use the calculation of the buildings' projected annual energy cost from EA Credit 1, or you can use a survey from the Department of Energy database.[1] Renewable energy must provide 2.5 percent of all annual energy consumed to achieve one point, 7.5 percent to achieve two points, and 12.5 to achieve all three points.

Sweetwater Sound investigated erecting wind turbines for this purpose. A wind turbine that would provide 1 megawatt of energy, enough to supply all of the energy needed for Sweetwater Sound, would be approximately 300 feet tall and would cost approximately one million dollars. The electricity needed to operate the Sweetwater Sound building costs approximately 140,000 dollars per year, putting the return on investment at just over seven years. They did not achieve any of these points yet, but are considering this investment for the future.

EA Credit 3: Enhanced Commissioning (1 Point)

The intention of this credit is to expand the amount of commissioning that is accomplished. It helps to underscore the value that commissioning brings to the project. This credit brings the CxA into the design phase of the project and continues their work beyond verifying systems performance. In order to achieve this credit, the following tasks must be added to the requirements in EA Prerequisite 1. First, the CxA must conduct a design review of the OPR, the BOD, and all design documents at some point before the mid-construction document review required in Prerequisite 1. In addition, during this document review, the CxA should also look at all review comments that had been included in the design submission. Second, at the same time as the A/E reviews, the CxA should review all of the commissioned systems' contractor submittals to make sure they comply with the OPR and BOD. Third, a systems manual should be developed for the operating staff. (This step does not have to be directly performed by the CxA.) Fourth, ensure that the operating staff and the occupants of the building receive the training they need to have in order to operate commissioned building systems to their maximum efficiency level. (This step doesn't have to be performed directly by the CxA.) Finally, the CxA must review the building operation with the operations and management staff within ten months of substantial completion. Any unresolved commissioning issues must have a written plan for resolving them.

Sweetwater Sound expects to receive this credit. They have contracted with a CxA for the extra involvement required, especially with owner training and follow-through. The additional expense related to this credit was 23,000 dollars.

EA Credit 4: Enhanced Refrigerant Management (1 Point)

The intention of this credit is to comply with the Montreal Protocol, to stop depleting the ozone layer in our Earth's atmosphere, and thereby, to stop contributing to global warming. The Montreal Protocol is an agreement signed by countries in 1987 to eliminate the use of the man-made chemicals, CFCs, halons, and other industrial chemicals that destroy the ozone layer. Without the Montreal Protocol agreement, the level of ozone depletion in our stratosphere would be at least five times greater than it is today, and the radiation levels of ultraviolet-B would have doubled. We are expected to achieve pre-1980 levels of CFC concentration by the year 2050.[2]

There are two ways to achieve this credit. The first is very simple: Do not use any refrigerants. The second method can also be simple. If you are using refrigerants or installing a fire-suppression system, don't use those that emit compounds that deplete the ozone layer. In this section, EA Credit 4, of the LEED-NC 2.2 Reference Manual, there is a formula for figuring out the maximum ozone-depleting compounds that are allowed. HVAC units and other equipment that contains less than one-half pound of refrigerant are not included in the requirements of this credit.

Sweetwater Sound did not achieve this credit. Their ice storage system saves money in operating costs because it works at night to make ice when electricity is cheaper—it costs about half as much as electricity costs during daytime peak hours. But this system actually takes more energy to operate. Ice is made at night and then is used during the day to cool the building until the ice has been completely used. At that point, the system switches to electricity generated cooling. There are four chillers that make the ice at Sweetwater Sound, and glycol fluid runs inside of a closed loop system just between the chillers. Glycol fluid contains HCFC R123. When the LEED® Accredited Professional (AP) asked why glycol was used, he was told that it is the fluid that operates the chillers most efficiently.

This is one of those places where we end up working at cross purposes between our intention, the electric company, and the environment. The electric company gains by this kind of system because it reduces loads during peak hours of power use, so it benefits them to sell electricity for half as much during the night time hours. The electric company wins by increasing its off-peak hour sales; the owner wins because operating costs are lessened; but the environment (and society) loses because more energy (which is still predominantly carbon-based) is being used to produce the cooling required, thereby releasing more carbon pollutants into the atmosphere. This is an ethical issue that has to be addressed with sustainability as the goal.

Chillers at Sweetwater Sound. Photograph Taken by the Author.

EA Credit 5: Measurement & Verification (1 Point)

The intention of this credit is to ensure that the building continues to operate at its original level of energy efficiency. In order to achieve this credit, a measurement and verification plan should be developed and implemented that covers at least one year of occupancy following construction. The plan should be consistent with "Option B or D" in the *International Performance Measurement & Verification Protocol Volume III: Concepts and Options for Determining Energy Savings in New Construction, April, 2003.*[3] Install metering equipment that measures energy use and compare the actual use with the predicted use to determine whether it is operating at the expected level of energy efficiency.

Sweetwater Sound did achieve this credit. They have installed the necessary metering equipment and are monitoring energy use to increase long-term efficiencies.

EA Credit 6: Green Power (1 Point)

The intention of this credit is to encourage us to develop and use renewable energy that is available from the grid. In order to gain this credit, a minimum of 35 percent of the electricity supplied to the building must come from renewable energy sources. The contract with the providing utility must be for the duration of at least two years.

Sweetwater Sound has a two-year contract with Northeastern REMC, a Touchstone Energy Provider for Northeast Indiana, to purchase renewable energy at a cost premium that supplies 35 percent of the buildings electricity consumption. They did achieve this credit.

Summary

This chapter is concerned with orienting the building so that it is able to capture and store as much renewable energy as possible. It discusses the problems encountered in trying to design larger buildings with the capacity for passive cooling. A larger building is described that has been able to accomplish passive, convection cooling using automated system controls—it is the Environmental Studies building at Oberlin College. Conventional HVAC systems heat and cool air, and air must be continually re-heated and re-cooled because it cannot maintain a temperature for any length of time.

We follow our commercial project through the LEED prerequisites and credits in energy and atmosphere, noting what they were able to achieve and what they weren't, with an explanation of why or why not. The three prerequisites involve commissioning of the building energy systems, providing minimum energy performance that complies with ASHRAE Standard 90.1, and using no CFC-based refrigerants in any new HVAC systems. Ten of the 17 possible credits concern optimizing energy performance, 3 concern on-site renewable energy, 1 concerns enhanced commissioning, another concerns enhanced refrigerant management that complies with the Montreal Protocol, one concerns measuring and verifying energy use, and the final credit concerns using renewable energy from the grid.

Questions and/or Assignments

1. Do some research on the building in which you work or go to school. Find out if it has an energy recovery system. Whether it does or doesn't, figure out how much it would (or does) save in both energy and money each year.
2. Find out what kind of HVAC system heats and cools the building in which you work or go to school. How efficient is it? Does it heat and cool air or mass?
3. What Standard is concerned with minimum levels of energy efficiency in building's HVAC systems?
4. Where will you find the LEED® worksheets that will assist in documentation of EA Prerequisite 2?
5. What is the intention of EA Prerequisite 3? And why is it important?
6. In EA Credit 1: Optimize Energy Performance, there are three methods you can use to accomplish this credit. Write a brief description of what they are.
7. In EA Credit 2: On-Site Renewable Energy, what percentage of the building's yearly energy use must be supplied by renewable energy sources in order to achieve from one to three points?
8. In EA Credit 3: Enhanced Commissioning, what steps in addition to verifying systems performance must be done by the commissioning agent?
9. The EA Credit 4: Enhanced Refrigerant Management refers to the Montreal Protocol. Explain what the Montreal Protocol is and why it was established.
10. What is the purpose of EA Credit 5: Measurement and Verification? And what is the name of the entity whose protocol must be followed in the plan that is created?
11. Explain EA Credit 6: Green Power. What percent of the electricity must be supplied by green power in order to achieve this credit?

Notes

1. This survey can be found on the Department of Energy website: http://www.doe.gov. Type the words "Commercial Buildings Energy Consumption Survey" into the search box to bring it up.
2. See the United Nations Environment Programme website: http://ozone.unep.org.
3. You can download this Protocol from http://www.eere.energy.gov/buildings/info/documents/pdfs/29564.pdf.

9

■ ■ ■

Materials and Resources

Abstract

This chapter is concerned with materials and resources. Embodied energy is the most important concept that we must understand about materials—it is the amount of energy that must be used in order to acquire and process raw materials. When the embodied energy in a material is high, it means that the impact on the environment is also high.

We follow our commercial project through the LEED prerequisite and credits in materials and resources, noting what they were able to achieve and what they weren't, with an explanation of why or why not.

The most important materials and resources concept is that of embodied energy. Once we understand the concept, the extension of the concept into rules and actions becomes transparent. Embodied energy is the amount of energy that must be used to acquire and process raw materials. In addition to the energy needed to extract the raw materials, we have to include the energy used to manufacture, transport, and install them. There is a direct relationship between the embodied energy quotient and the environmental impact of a material. When the embodied energy in a material is high, it means that the impact on the environment is also high. The embodied energy quotient of a material should also include the projected time the product will be in use. This way we get a truer figure representing embodied energy because it takes durability into account. Some products have a high embodied energy quotient when extracted raw materials are used to produce them, like steel or aluminum. But when the production process uses recycled material instead of raw, those same products have a very low embodied energy quotient. In the following table, look at the large differences in the embodied energy quotients of steel and aluminum products made with raw materials, and those made with recycled materials. You can also find the embodied energy of some common construction materials.

	Embodied Energy	
Material	**MJ/kg***	**MJ/m³****
Aggregate	0.10	150
Straw bale	0.24	31
Stone (local)	0.79	2,030
Concrete block	0.94	2,350
Concrete (30 Mpa)	1.30	3,180
Concrete precast	2.00	2,780
Lumber	2.50	1,380
Brick	2.50	5,170
Cellulose insulation	3.30	112
Gypsum wallboard	6.10	5,890
Particle board	8.00	4,400
Aluminum (recycled)	8.10	21,870
Steel (recycled)	8.90	37,210
Shingles (asphalt)	9.00	4,930
Plywood	10.40	5,720
Glass	15.90	37,550
Fiberglass insulation	30.30	970
Steel	32.00	251,200
Zinc	51.00	371,280
Brass	62.00	519,560
PVC	70.00	93,620
Copper	70.60	631,164
Paint	93.30	117,500
Linoleum	116.00	150,930
Polystyrene insulation	117.00	3,770
Carpet (synthetic)	148.00	84,900
Aluminum (recycled)	227.00	515,700

*Megajoules per kilogram of nonrenewable energy per unit of material.

**Megajoules per cu. ft. of nonrenewable material.

Source: The Canadian Architect website: http://www. canadianarchitect.com.

USGBC LEED-NC MATERIALS AND RESOURCES, 1 PREREQUISITE AND 13 POSSIBLE POINTS

MR Prerequisite 1: Storage and Collection of Recyclables (Required)

The intention of this prerequisite is to reduce the landfilled waste that is generated by the occupants of the building. To achieve this prerequisite, an area that is easily accessible to building occupants for the collection of recyclable materials must be established. All corrugated cardboard, plastic, metal, glass, and paper must have collection bins provided.

Sweetwater Sound did achieve this prerequisite by providing recycling bins that are easy to access by all building occupants.

MR Credit 1.1: Building Reuse—Maintain 75 Percent of Existing Walls, Floors, and Roof (1 Point)

The intention of this credit is to conserve material resources by extending the life of existing buildings. In order to achieve the credit, at least 75 percent of the surface area of the existing structure must be retained. If the square footage of the new construction being added to the existing structure is more than twice the size of the existing building, this credit cannot be achieved.

Sweetwater Sound did not achieve this credit. Although they did reuse an existing building, the new construction addition is more than twice the square footage of the reused building. The original building was a one-story structure with a 38,000-square-foot footprint, and is home to most of the sales and administrative offices. The new structure provides an additional 117,300 square feet and houses the warehouse, auditorium, recording studios, and a main concourse with retail store, restaurant, and other employee services. In the rendering below, the rectangular shaped structure in the upper left part of the building is the original structure that Sweetwater purchased.

MR Credit 1.2: Building Reuse—Maintain 95 Percent of Existing Walls, Floors, and Roof (1 Point in Addition to MR Credit 1.1)

This credit is an extension of MR Credit 1.1, simply adding 20 percent to the total square foot of surface area retained and reused in the existing building. And of course, the same caveat applies as the one mentioned above.

Sweetwater Sound Rendering.   *Sweetwater Sound/MSKTD. Reproduced by Permission.*

MR Credit 1.3: Building Reuse—Maintain 50 Percent of Interior Nonstructural Elements (1 Point)

This credit has the same intention of the two prior credits—to extend the life of existing buildings and thereby conserve material resources. In order to achieve this credit, at least 50 percent of all existing building elements that are not structural must be maintained and reused. This applies to items like floor coverings, ceiling systems, and interior walls and doors. But again this credit does not apply if the square footage of the new construction is more than two times the size of the existing building being reused.

Sweetwater Sound did not apply for this credit, because the new construction is more than two times the size of the reused building.

MR Credit 2.1: Construction Waste Management—Divert 50 Percent from Disposal (1 Point)

The intention of this credit is to dramatically reduce the amount of construction "waste" that goes to the landfill. Nearly everything that gets thrown into a dumpster on a jobsite that would be headed for the local landfill is a material that can be used for another purpose. Cardboard, metals, plastics, asphalt from parking lots to shingles, concrete demolition material, drywall remnants, wood remnants, and so on can all be recycled. All of these materials can be used as the recycled resource material for other products. In order to achieve this credit, a plan to manage the construction waste on the jobsite must be developed and put into action. The plan must detail the materials to be salvaged and whether they will be sorted at the jobsite, or co-mingled and sorted elsewhere. This credit excludes excavated soil and debris cleared from the land. You can choose to calculate using the weight of the material, or the volume, but the key is to be consistent with that choice on all of the material calculations.

Sweetwater Sound did achieve this credit. They recycled all steel studs, block, conduit, and drywall that were removed from the interior of the existing building. During construction, they recycled all cardboard, plastics, drywall, wood, and steel. The asphalt and concrete that had to be removed in order for the new building to be constructed were also recycled.

MR Credit 2.2: Construction Waste Management—Divert 75 Percent from Disposal (1 Point in Addition to MR Credit 2.1)

The intention of this credit is the same as in MR Credit 2.1. This just asks that an additional 25 percent of construction waste be recycled, thereby diverting it from the landfill.

Sweetwater Sound did achieve this credit as well by diverting an astounding 98 percent of construction waste from the landfill.

MR Credit 3.1: Materials Reuse—5 Percent (1 Point)

The intention of this credit is to encourage the reuse of any building material that is considered to be permanently installed in the project—things like beams, posts, cabinetry, doors, and the alike. If furniture is included in this credit, it

Removed Asphalt, Ground, and Ready to be Reused On Site.   *Sweetwater Sound/John Hopkins. Reproduced by Permission.*

must be used consistently throughout MR Credits 3–7. The calculation is based on cost. The sum of the salvaged materials must make up at least 5 percent of the total value of all project materials.

Sweetwater Sound did not achieve this credit. The value of reused project materials did not come anywhere near to 5 percent of the total materials value.

MR Credit 3.2: Materials Reuse—10 Percent (1 Point in Addition to MR Credit 3.1)

Again, the intention of this credit is exactly the same as MR Credit 3.1. To attain the credit, the materials being reused must constitute at least 10 percent of the total value of all project materials. The calculation base used should be the same as above.

Sweetwater Sound did not achieve this credit.

MR Credit 4.1: Recycled Content—10 Percent (Postconsumer + $1/2$ Preconsumer) (1 Point)

The intention of this credit is to expand the use of building products that are made with recycled material content. The International Organization of Standards document, *ISO 14021—Environmental labels and declarations—Self-declared environmental claims (Type II environmental labeling)*, is the one that LEED® relies on to establish the definition of "recycled content."[1] Preconsumer material, as its name implies, has never been used in the consumer market—it is defined as salvaged

All Steel in Project Is Recycled Material from a Local Company, Steel Dynamics.   *Sweetwater Sound/John Hopkins. Reproduced by Permission.*

material that is generated through the process of manufacturing. Postconsumer material is something that has completed its life cycle as a consumer item, and has been diverted from the landfill. This would be things like plastic bottles that have been turned into decking members, or jeans that have been turned into insulation. Products containing recycled content may contain some preconsumer material, some postconsumer material, or some of both. This credit specifies that only permanently installed materials can be included in the calculation, and that if furniture is included, it again, must be consistently applied through MR Credits 3–7. The calculation rule for this credit states that one-half of the preconsumer recycled content plus the postconsumer recycled content in materials must be equal to at least 10 percent of the total value of all project materials (based on cost).

Sweetwater Sound easily achieved this credit, at no additional cost to the project. Thirty-one percent of all finish materials are made with recycled content. All steel used on this project contained 80–90 percent recycled material. All carpet, drywall, ceiling tile, terrazzo tile, and concrete had recycled content.

MR Credit 4.2: Recycled Content—20 Percent (Postconsumer + $^{1}/_{2}$ Preconsumer) (1 Point in Addition to MR Credit 4.1)

The intention of this credit is exactly the same as MR Credit 4.1. It awards another point for achieving an additional 10 percent recycled content materials beyond MR Credit 4.1.

Sweetwater Sound did attain this credit in addition to the last one.

Terrazzo Tile/Recycled Content on Main Concourse of Sweetwater Sound. *Sandstone on Wall Came from a Nearby Quarry. Photograph Taken by the Author.*

MR Credit 5.1: Regional Materials—10 Percent Extracted, Processed, and Manufactured Regionally (1 Point)

The intention of this credit is to encourage the use of building materials and products that are manufactured or extracted within a 500-mile radius of the job site. This credit reduces the need to move materials over great distances, thereby reducing the environmental consequences wreaked by transportation. The calculation for this credit is again based on cost. At least 10 percent of the total value of materials must originate, whether extracted or manufactured, regionally (defined as within 500 miles from the building site). Only the percentage of the material that is derived regionally can be included in the calculation. Again, this credit only applies to those materials that are permanently installed, and if furniture is included in the calculation, it must be consistently used through MR Credits 3–7.

Sweetwater Sound did achieve this credit, as 44 percent of all finish materials came from within a 500-mile radius.

MR Credit 5.2: Regional Materials—20 Percent Extracted, Processed, and Manufactured Regionally (1 Point in Addition to MR Credit 5.1)

The intention of this credit is exactly the same as MR Credit 5.1, and this just asks the percentage of regional materials be increased to 20 percent.

Sweetwater Sound achieved this additional credit as well, with no extra cost associated.

MR Credit 6: Rapidly Renewable Materials (1 Point)

The intention of this credit is to encourage the use of raw materials that take ten years or less to regrow following harvest, and to discourage the use of raw materials that take longer than ten years to regrow. The calculation is again based on cost. To achieve this credit, 2.5 percent of the total value of all building materials must be rapidly renewable. Consider materials like linoleum, cork, wheatboard, wool, and cotton insulation in fulfilling this credit.

Sweetwater Sound did achieve this credit by installing bamboo and linoleum equal to 2.5 percent of the total value of building materials.

MR Credit 7: Certified Wood (1 Point)

The intention of this credit is to encourage forest management practices that are responsible to the environment. In order to achieve this credit, 50 percent of all wood and wood products that are permanently installed in the building must be Forest Stewardship Council (FSC) certified. This would include things like structural wood members, wood doors, flooring, and finishes, and can include furniture as long as it has been included consistently in MR Credits 3–7. To locate FSC-certified building products, go to their website http://www.fscus.org/.

Bamboo Cladding on Main Concourse Wall, and Bamboo Flooring in Retail Sales Area. Photograph Taken by the Author.

Sweetwater Sound did achieve this credit by using a minimum of 50 percent of FSC-certified wood-based materials. Structural framing, flooring, sub-flooring, wood doors, and finishes in the building are all FSC-certified wood materials. The FSC-certified wood cost 2 percent more than regular wood.

Summary

This chapter is concerned with materials and resources. Embodied energy is the most important concept that we must understand about materials. Embodied energy is the amount of energy that must be used in order to acquire and process raw materials. Acquisition and processing raw materials includes extraction, manufacturing, transportation, installation, and the projected amount of time the material will be in use. There is a direct relationship between the embodied energy quotient and the environmental impact of a material. When the embodied energy in a material is high, it means that the impact on the environment is also high.

We follow our commercial project through the LEED prerequisite and credits in materials and resources, noting what they were able to achieve and what they weren't, with an explanation of why or why not. The only prerequisite requires that an easily accessible space be established for the storage and collection of recyclables by building occupants. Of the 13 possible credits in this section, our commercial company did not achieve those based on a calculation in building reuse and materials reuse because even though the existing building was fully reused, the new construction is more than twice the size of the existing building. The project easily achieved credits in the sections on construction waste management, recycled content, regional materials, rapidly renewable materials, and certified wood.

Questions and/or Assignments

1. Write a brief paragraph explaining the embodied energy of something simple that you use everyday, like a pencil or a glass of milk.
2. In MR Prerequisite 1: Storage and Collection of Recyclables, what are the recyclable materials that must have collection bins provided for them in order to achieve this prerequisite?
3. In MR Credit 1, 1.2, and 1.3: Building Reuse, a certain percentage of the existing building and the interior nonstructural elements must be reused. If the addition is a 25,000 sq. ft. warehouse being added to an existing 12,000 sq. ft. building, and the entire existing building is being reused, will this credit point be achieved?
4. In MR Credit 2.1: Construction Waste Management, what percent of construction waste must be diverted from the landfill?
5. Regarding MR Credit 2.2: Construction Waste Management, several tons of clay were excavated and recycled from a jobsite in order to create a space for a rain garden. Can this clay be included in the calculation for this credit? If so, should it be calculated by weight or volume?
6. MR Credits 3.1 and 3.2: Materials Reuse, specify that at least 5 percent of permanently installed materials must be reused for the first point and 10 percent to achieve the

second point. If Company XYZ reused existing materials that would cost 80,000 dollars, and the total cost of materials for the project was 250,000 dollars, were they able to achieve either of these credits?

7. What is the difference between preconsumer and postconsumer recycled content? Give an example of each.

8. What document does LEED MR Credits 4.1 and 4.2: Recycled Content rely on to establish the definition of "recycled content?" What is the calculation rule for MR Credits 4.1 and 4.2?

9. A building owner wants to achieve MR Credits 5.1 and 5.2: Regional Materials, and bamboo cabinetry and furniture being used are all manufactured within a 500-mile radius; however, the bamboo itself comes from beyond that limit. How would you calculate this?

10. In order to be considered a rapidly renewable material, in how many years must a raw material be able to replenish or regrow itself? Give some examples of rapidly renewable materials. What percentage of all building materials must be rapidly renewable in order to achieve MR Credit 6?

11. Regarding MR Credit 7: Certified Wood, what percentage of wood and wood products must be FSC certified to attain this point? Can furniture be included in this calculation? And if so, what is the rule regarding furniture?

12. Go to http://www.fscus.org to answer this question. Explain what the Forest Stewardship Council does and write a brief description of their mission.

Notes

1. See the International Organization of Standards at http://www.iso.org/iso/home.htm.

10

■ ■ ■

Indoor Quality—Air, Light, and Views

Abstract

This chapter is concerned with indoor quality of air, daylight, and occupant views. We follow our commercial project through the LEED prerequisites and credits in indoor environmental quality, noting what they were able to achieve and what they weren't, with an explanation of why or why not. This section of LEED is concerned with the health and comfort of the human beings who work in the built environment.

USGBC LEED-NC INDOOR ENVIRONMENTAL QUALITY, TWO PREREQUISITES AND 15 POSSIBLE POINTS

EQ Prerequisite 1: Minimum IAQ Performance (Required)

The intention of this prerequisite is to ensure that the quality of the air inside of a building complies with a minimum standard established in ASHRAE Standard 62.1-2004, Sections 4–7, "Ventilation for Acceptable Indoor Air Quality." If the local code is stricter than this ASHRAE standard, the local code must be followed. Buildings that have operable windows, or are otherwise ventilated naturally, must comply with ASHRAE Standard 62.1-2004, paragraph 5.1.

Sweetwater Sound met this prerequisite by complying with ASHRAE 62.1-2004.

EQ Prerequisite 2: Environmental Tobacco Smoke (ETS) Control (Required)

The intention of this prerequisite is to protect building occupants from air contaminated with tobacco smoke. This prerequisite can be achieved in a couple of different ways. The first method includes not allowing smoking inside of the building at all. In addition, designated smoking areas outside of the building must be located a minimum of 25 feet from any locations where outside air could enter the building. This should include things like doors, operable windows, and all exterior air intake vents.

A second method would allow smoking inside of the building, but only in a designated smoking room from which all ETS is completely captured and exhausted, and in which a negative air pressure is established in the room so that when the door is opened to the inside of the building, no contamination of the nonsmoking area occurs. The performance of the room must also be tested and shown to be working as it is designed to work, with no contamination of the air outside of the smoking room. Again, just as in the first case, any designated smoking areas located outside of the building must be placed at least 25 feet from any place where outside air could enter the building.

The third method of complying with this prerequisite is limited to residential buildings. In this case, smoking is only allowed inside of a private residence, and is not allowed in any common areas. And again, if there are designated smoking areas outside of the building, they must be placed at least 25 feet from any operable windows that open into common areas, entry doors, and air intake vents. In addition, all penetrations in the individual residential units and their adjacent vertical chases must be sealed to eliminate ETS contamination between units. Residential units must be proved to have sealing that complies with both of the following standards: The first is ANSI/ASTM-E779-03, "Standard Test Method for Determining Air Leakage Rate by Fan Pressurization" and the second is Chapter 4 of the Residential Manual for Compliance with California's 2001 Energy Efficiency Standards.

Sweetwater Sound did comply with this prerequisite. They do not allow smoking anywhere on their campus, unless the individual is inside of their own vehicle with the windows rolled up. (Most employees at Sweetwater Sound who did smoke when they began working for the company have stopped.) A new sign at the entrance to the building advises guests and employees that Sweetwater is a smoke-free campus.

EQ Credit 1: Outdoor Air Delivery Monitoring (1 Point)

The intention of this credit is to ensure the safety of building occupants by monitoring ventilation systems to make certain they are working properly. In order to achieve this point, a permanent monitoring system must be installed that will set an alarm off if the ventilation system varies by 10 percent or more. Whether the space is ventilated using mechanical or natural methods, carbon dioxide monitors must be installed, between 3 and 6 feet from the floor, in all densely occupied rooms (25 or more people per 1,000 square feet of space). In spaces that are not densely occupied, equipment that measures outdoor air should be installed according to specifications defined in ASHRAE 62.1-2004.

Sweetwater Sound has the appropriate equipment installed and achieved this credit.

EQ Credit 2: Increased Ventilation (1 Point)

The intention of this credit is to increase the amount of outdoor air ventilated into interior spaces by at least a 30-percent improvement over the minimum requirements of ASHRAE Standard 62.1-2004. To achieve this credit in space that is occupied and mechanically ventilated, improve by 30 percent the minimum requirements of outdoor air ventilation determined in EQ Prerequisite 1. To achieve this credit in occupied, naturally ventilated space, refer to the Carbon Trust "Good Practice Guide 237" of 1998.[1] This guide assists in thinking through the process of ventilating spaces naturally. Also, refer to the Chartered Institution of Building Services Engineers (CIBSE) Applications Manual 10:2005, "Natural Ventilation in Non-Domestic Buildings," Figure 1.18, and follow the flow diagram to develop a natural ventilation strategy. This credit requires heavy documentation. You can document by using diagrams and calculations or by using multizone modeling software. Be sure to avoid additional energy expense and use by incorporating heat recovery systems to precondition the incoming outdoor air.

Sweetwater Sound did not apply for this credit.

EQ Credit 3.1: Construction IAQ Management Plan—During Construction (1 Point)

The intention of this credit is to make sure that the quality of the air during construction is not detrimental to the people working in the building. In order to attain this credit, an Indoor Air Quality Management Plan must be developed and implemented for all stages of the building process prior to its being occupied. There are three aspects of this credit that must be attended to in the plan and its implementation: (1) All absorptive materials stored on site must be kept dry; (2) all air handlers that are permanently installed must be fitted out with filters with a Minimum Efficiency Reporting Value (MERV) of 8 at return air grilles (as referenced by ASHRAE Standard 52.2-1999), and the filters must be replaced just before the building is occupied; and (3) the air during construction must meet the requirements of the Sheet Metal and Air Conditioning National Contractors Association (SMACNA) IAQ Guidelines for Occupied Buildings under Construction, 1995, Chapter 3.

Sweetwater Sound achieved this credit by complying with all three requirements.

EQ Credit 3.2: Construction IAQ Management Plan—Before Occupancy (1 Point)

The intention of this credit is to ensure the quality of air is problem-free before the building is occupied. This credit can be achieved by designing and implementing an IAQ Management Plan in one of the following two ways. One method is by conducting a complete flush-out of the building after construction is completed and before it is occupied. The second method requires that the IAQ be tested to ensure that concentration levels of various contaminants do not exceed the guidelines established by the U.S. Environmental Protection Agency "Compendium of Methods for the Determination of Air Pollutants in Indoor Air."[2]

Air Ducts Covered During Construction. *Sweetwater Sound/John Hopkins. Reproduced by Permission.*

Sweetwater Sound achieved this credit by conducting a complete building flush (14,000 cubic feet per square foot of floor space) after construction was fully completed and before the building was occupied.

EQ Credit 4.1: Low-Emitting Materials—Adhesives and Sealants (1 Point)

The intention of this credit is to ensure the high quality of indoor air by reducing contaminants emitted from adhesives and sealants that are installed inside of the weatherproofing system of the building. In order to achieve this credit, the volatile organic compound (VOC) limits established in the South Coast Air Quality Management District (SCAQMD) Rule 1168 (rule amendment dated 1/7/2005) for adhesives and sealants must be complied with. In addition, the Green Seal Standard for Commercial Adhesives GS-36 requirements for aerosol adhesives, dated 10/19/2000, must be complied with.

Sweetwater Sound did achieve this credit.

EQ Credit 4.2: Low-Emitting Materials—Paints and Coatings (1 Point)

The intention of this credit is to ensure the high quality of indoor air by reducing contaminants emitted from paints and coatings that are installed inside of the weatherproofing system of the building. In order to achieve this credit, VOC levels of interior paints and coatings on walls and ceilings must comply with the Green Seal Standard GS-11, Paints, First Edition, May 20, 1993; VOC levels of anticorrosive and antirust paints on interior metals must comply with the Green Seal Standard GC-03, Anti-Corrosive Paints, Second Edition, January 7, 1997; and VOC levels of clear wood finishes and stains must comply with the SCAQMD Rule 1113, Architectural Coatings, from January 1, 2004.

Sweetwater Sound did achieve this credit. All paints and coatings were specified to ensure compliance with the Green Seal Standard and SCAQMD Rule 1113.

EQ Credit 4.3: Low-Emitting Materials—Carpet Systems (1 Point)

The intention of this credit is to ensure the high quality of indoor air by reducing contaminants emitted from carpet systems that are installed inside of the building. In order to achieve this credit, all carpet systems and cushion must comply with the requirements of the Carpet and Rug Institute's Green Label Plus program, and adhesives used in carpet installation must comply with EQ Credit 4.1.

Sweetwater Sound did achieve this credit.

Carpet System Being Installed at Sweetwater Sound.   *Sweetwater Sound/John Hopkins. Reproduced by Permission.*

EQ Credit 4.4: Low-Emitting Materials—Composite Wood and Agrifiber Products (1 Point)

The intention of this credit is to ensure the high quality of indoor air by reducing contaminants emitted from composite wood and agrifiber products that are installed inside of the weatherproofing system of the building. In order to achieve this credit, none of the composite wood and agrifiber products or adhesives may contain urea–formaldehyde resins. The products of concern here are things like plywood, wheatboard, particleboard, and door cores.

      Sweetwater Sound did achieve this credit.

EQ Credit 5: Indoor Chemical and Pollutant Source Control (1 Point)

The intention of this credit is to ensure the high quality of indoor air by controlling chemicals and pollutants from entering the building and/or from cross-contaminating occupied building areas. In order to achieve this credit, contamination must be prevented in three different ways. The first is to prevent it from "walking in" to the building from all outdoor entries. To accomplish this, either a 6-foot-long walk-off mat that gets changed weekly by a contracted mat service or 6 feet of grate/grill system that is permanently

Permanently Installed Grate/Grill System in Each Building Entry at Sweetwater Sound. Photograph Taken by the Author.

installed and can be cleaned underneath must be installed. The second is to prevent hazardous gasses or chemicals from leaving the rooms in which they are emitted. Rooms that contain any hazardous fumes must both completely contain and exhaust them. Negative air pressure must be created in the room, self-closing doors must be installed, and ceilings must be built as either deck to deck or hard lid. The third is concerned with mechanically ventilated buildings. The filters being used before the building is occupied must be with a MERV of 13 or greater and must filter all supply air, whether it is outside air or return air.

Sweetwater Sound did achieve this credit.

EQ Credit 6.1: Controllability of Systems—Lighting (1 Point)

The intention of this credit is to ensure that people have the ability to control the light they need for their work. In order to achieve this credit, a minimum of 90 percent of all people who occupy the building must be able to adjust the light they require for their work, and groups who work in spaces together must be able to adjust the light according to the task needs of the group.

Sweetwater Sound did achieve this credit by installing task lighting at every workstation. This added to the initial construction cost of the building by 10,000 dollars.

Task Lighting Is Installed at Each Workstation. Photograph Taken by the Author.

EQ Credit 6.2: Controllability of Systems—Thermal Comfort (1 Point)

The intention of this credit is to ensure that people have the ability to control the thermal comfort they need for their work. In order to achieve this credit, a minimum of 50 percent of those who occupy the building must be able to control thermal comfort for their own needs, and those who share space (e.g., classrooms and conference rooms) must also be able to adjust for thermal comfort. The ASHRAE Standard 55-2004 defines thermal comfort conditions and their control as including the ability to control one of the following: air and radiant temperature, air speed, and humidity.

Sweetwater Sound has applied for this credit. Accessible thermostats are installed on the walls of individual offices and group rooms.

EQ Credit 7.1: Thermal Comfort—Design (1 Point)

The intention of this credit is to ensure thermal comfort for those who occupy the building. In order to achieve this credit, the HVAC systems and the building envelope must comply with ASHRAE Standard 55-2004, Thermal Comfort Conditions for Human Occupancy. This must be documented according to Section 6.1.1.

Sweetwater Sound has complied with ASHRAE 55-2004 and did achieve this credit.

EQ Credit 7.2: Thermal Comfort—Verification (1 Point)

The intention of this credit is to ensure that the building HVAC systems continue to provide thermal comfort for building occupants over time. In order to achieve this credit, the building owners must agree to survey all building occupants, within 6–18 months post occupancy, to gage their satisfaction with the thermal comfort of their working environment. The responses must also be anonymous so that workers can give honest answers about their own comfort. If 20 percent or more of the survey results indicate thermal discomfort, the owners must agree to develop and implement a plan to correct the problem areas.

Sweetwater Sound did achieve this credit. See a copy of their survey at the end of this chapter.

EQ Credit 8.1: Daylight and Views—Daylight 75 Percent of Spaces (1 Point)

The intention of this credit is to ensure that those who work inside of the building have access to both daylight and views. There are three methods available to demonstrate that the requirements have been achieved. The first one is by using a calculation to prove that 75 percent of all occupied workspace has enough fenestration to provide a glazing factor of at least 2 percent. The second method is by conducting a computer simulation model. The model must show that 75 percent of all occupied workspace receives at least 25 horizontal foot-candles of daylighting. Furthermore, the modeling must prove these results under the following conditions: 30 inches above the floor, on the equinox, at noon, and under clear skies. The third method is by showing, using a 10-foot grid on the building's floor plans, that 75 percent of all occupied workspace receives at least 25 horizontal foot-candles of daylighting through measuring and recording indoor light measurements.

Sweetwater Sound achieved this credit. Any extra up-front cost is already counted in the Energy and Atmosphere credits.

EQ Credit 8.2: Daylight and Views—Views for 90 Percent of Spaces (1 Point)

The intention of this credit is again to ensure that those who work inside of the building have access to both daylight and views. In order to achieve this credit, 90 percent of all work areas must have a direct line of sight through windows located between 2.5 and 7.5 feet above the floor. Both plan and section drawings must show that the direct line of sight is available.

Sweetwater Sound did achieve this credit. No additional cost is associated with this credit.

Daylight and Views. Photograph Taken by the Author.

Daylight in Sweetwater Sound Warehouse.   *Sweetwater Sound/John Hopkins. Reproduced by Permission.*

Summary

This chapter is concerned with indoor quality of air, daylight, and occupant views. We follow our commercial project through the LEED prerequisites and credits in indoor environmental quality, noting what they were able to achieve and what they weren't, with an explanation of why or why not. This section of LEED is concerned with the health and comfort of the human beings who work in the space. The two prerequisites ensure that the quality of the indoor air is good and contains no problematic particulates. The first two credits are concerned with increased ventilation and monitoring the mechanical systems that deliver the air to make sure they are working properly. The next two credits are concerned with ensuring air quality during construction and before the building is occupied. The following four credits are concerned with installing materials that contain low-emitting VOCs. The last credit that concerns air quality is to ensure that chemicals and pollutants are prevented from contaminating building air. Two of the credits are meant to ensure that occupants have control of both lighting and thermal comfort for their own workspace. The following two credits are concerned with thermal comfort conditions for human occupancy in both design and verification. The last two credits are concerned with access for building occupants to daylight and views. An addendum to this chapter is the thermal comfort survey this company has used with employees.

Questions and/or Assignments

1. What is the Standard referred to in EQ Prerequisite 1: Minimum IAQ Performance?
2. What is the easiest way to satisfy the requirements for EQ Prerequisite 2: Environmental Tobacco Smoke Control?
3. In residential units, sealing must comply with two Standards. What are they?
4. In EQ Credit 1: Outdoor Air Delivery Monitoring, where must carbon dioxide monitors be placed in densely occupied rooms? What determines whether a room is densely occupied or not?
5. In EQ Credit 2: Increased Ventilation, what is the percentage of improvement over EQ Prerequisite 1 that would be necessary to achieve this point? In occupied, naturally ventilated space, what are the manuals that you should rely on for assistance? How can you document this?
6. What are the three aspects of EQ Credit 3.1: Construction IAQ Management Plan that must be attended to in the plan and implementation in order to attain this point?
7. During construction, what MERV value must the filters at return air grilles have?
8. Indoor air quality must meet specific guidelines during construction. What are they?
9. For EQ Credit 3.2: Construction IAQ Management Plan, the point can be achieved in one of two ways. What are they?
10. For EQ Credit 4.1: Low-Emitting Materials, the VOC limits were established by what organization, and must comply with what rule?
11. In EQ Credit 4.2: Low-Emitting Materials, what is the Standard to which interior paints and coatings must comply?
12. In EQ Credit 4.3: Low-Emitting Materials, what is the organization to which all installed carpet systems must comply?
13. In order to attain EQ Credit 4.4: Low-Emitting Materials, what is the ingredient that any composite wood and agrifiber products may not have?
14. In EQ Credit 5: Indoor Chemical and Pollutant Source Control, there are three ways in which contamination must be prevented. What are they?
15. In mechanically ventilated buildings, the filters being used before occupancy must be of what MERV value?
16. In EQ Credit 6.1: Controllability of Systems, what is the percentage of all people who occupy the building who must be able to control the light for their work?
17. In EQ Credit 6.2: Controllability of Systems, what percentage of building occupants who must be able to control thermal comfort for their own needs?
18. In EQ Credit 7.1: Thermal Comfort, what is the Standard that defines thermal comfort conditions?
19. In EQ Credit 7.2: Thermal Comfort, building occupants must be surveyed to discover satisfaction levels with the thermal comfort of their working environment. When must this be done? If 20 percent or more of the results indicate discomfort, what must the building owners agree to do?
20. In EQ Credit 8.1: Daylight and Views—Daylight 75 Percent of Spaces—explain one of the ways of documenting for this point to be attained.
21. In EQ Credit 8.2: Daylight and Views—Views for 90 Percent of Spaces—how far above the floor must the windows be located in order to provide a direct line of sight?

Notes

1. See http://www.carbontrust.co.uk/documents for details.
2. http://www.epa.gov/nrmrl/pubs/625r96010/iocompen.pdf for complete downloadable guidelines.

Sweetwater Thermal Comfort Survey

This survey is intended to provide an assessment of the thermal comfort provided by this building to its employees. Answers to these survey questions provide an indication as to the performance of the building's heating, ventilation, and air-conditioning systems while providing direction for making improvements to systems in an attempt to provide a continual comfortable environment for building occupants.

This survey is divided into four sections:

Section 1–Background information.
Section 2–Assessment of the current conditions in your space.
Section 3–Assessment of the conditions in your space over the course of the winter months.
Section 4–Assessment of the conditions in your space over the course of the summer months.

Thank you for your participation!

Section 1
Background Information

How many years have you worked in this building?

☐   Less than 1 year   ☐   1–2 years

☐   3–5 years   ☐   More than 5 years

On which floor is your office located?

☐   First floor   ☐   Second floor

In which direction does your office face?

☐   East   ☐   West   ☐   Northwest corner

☐   Northeast corner   ☐   Southwest corner   ☐   Southeast corner

Which of the following do you use to adjust or control in your office environment? (check any that apply)

☐ Window blinds or shades   ☐ Thermostat   ☐ Portable heater

☐ Room air-conditioning unit   ☐ Portable fan

☐ Ceiling fan   ☐ Adjustable air vents

☐ Windows   ☐ Other

If other please describe _____

(*continued*)

Section 2
Current Thermal Comfort
The following questions refer to the current conditions / comfort level you perceive at the time you are completing this survey.

Date: _____

Time: _____

What are the seasonal conditions outside?

☐   Winter   ☐   Spring

☐   Summer   ☐   Fall

What is the approximate temperature outside today? (degrees fahrenheit)

_____ °F

How would you describe the weather outside today?

☐   Clear skies / sunny   ☐   Overcast

☐   Partly cloudy

What is your current thermal comfort:

1. ☐ Hot
2. ☐ Warm
3. ☐ Slightly warm
4. ☐ Neutral
5. ☐ Slightly cool
6. ☐ Cool
7. ☐ Cold

How satisfied are you with the temperature in your office today?

Very satisfied   ☐ ☐ ☐ ☐ ☐ ☐ ☐   Very dissatisfied

If you are dissatisfied, how would you best describe the source of your discomfort? (check all that apply)

☐ Air movement too high   ☐ Air movement too low   ☐ Incoming sun

☐ Drafts from windows   ☐ Drafts from vents   ☐ Humidity level too high/low

☐ Hot/cold surrounding surfaces (floor, ceiling, walls, or windows)

☐ Heating/cooling system does not respond quickly enough to the thermostat

☐ Other. Please describe: _____

(*continued*)

(continued)

Are any of the following currently operating in your office?

☐ Computers / lap tops   ☐ Lighting   ☐ Other.
                                                Please describe:_____

☐ Copier / Fax machine   ☐ Dishwasher

Clothing: Please place a check by the articles of clothing that you are wearing (this is an indication as to the comfort level of your interior space):

Top	**Bottom**
☐ Short sleeve shirt	☐ Trousers
☐ Long sleeve shirt	☐ Knee—length skirt
☐ Sweater vest	☐ Walking shorts
☐ Suit vest	☐ Overalls
☐ Long sleeve sweater	☐ Jeans
☐ Long sleeve sweatshirt	☐ Athletic sweat pants
☐ T-shirt	☐ Ankle—length skirt
☐ Thermal underwear top	☐ Thermal underwear bottoms

How would you describe your activity level just prior to completing this survey?

☐ Seated quiet   ☐ Standing relaxed   ☐ Light activity, standing

☐ Medium activity, standing   ☐ High activity

Section 3
Seasonal Comfort, Winter
The following questions refer to your general perception of thermal comfort in your office / retail space throughout the winter months.

In the winter months, how satisfied are you with the temperature in your office?

Very satisfied   ☐ ☐ ☐ ☐ ☐ ☐ ☐   Very dissatisfied

If you are dissatisfied, would you describe the temperature as too hot or too cold?

☐   Too hot   ☐   Too cold

(continued)

If you are dissatisfied, how would you best describe the source of your discomfort? (check all that apply)

☐ Air movement too high   ☐ Air movement too low   ☐ Incoming sun

☐ Drafts from windows   ☐ Drafts from vents   ☐ Humidity level too high/low

☐ Hot/cold surrounding surfaces (floor, ceiling, walls, or windows)

☐ Heating/cooling system does not respond quickly enough to the thermostat

☐ Uneven temperature (some parts always hot while others always cold)

☐ Other. Please describe: _____

Are you satisfied with the acoustic performance of your office? (i.e., is your office too noisy from fans, vibrations, or ambient noise?)

Generally quiet ☐ ☐ ☐ ☐ ☐ ☐ ☐ Too noisy

Section 4
Seasonal Comfort, Summer
The following questions refer to your general perception of thermal comfort in your office / retail space throughout the summer months.

In the summer months, how satisfied are you with the temperature in your office?

Very satisfied   ☐ ☐ ☐ ☐ ☐ ☐ ☐ Very dissatisfied

If you are dissatisfied, would you describe the temperature as too hot or too cold?

☐   Too hot   ☐   Too cold

If you are dissatisfied, how would you best describe the source of your discomfort? (check all that apply)

☐ Air movement too high   ☐ Air movement too low   ☐ Incoming sun

☐ Drafts from windows   ☐ Drafts from vents   ☐ Humidity level too high/low

☐ Hot/cold surrounding surfaces (floor, ceiling, walls, or windows)

☐ Heating/cooling system does not respond quickly enough to the thermostat

☐ Uneven temperature (some parts always hot while others always cold)

☐ Other. Please describe: _____

11

■ ■ ■

Innovation and Design

Abstract

Chapter 11 is concerned with the innovation and design LEED credits for which our commercial project has applied. This project has gone beyond the LEED requirements in a number of areas, but four credits are the maximum allowed in this segment. The final credit in this section is satisfied by the employment of a LEED Accredited Professional to facilitate the process of registration, application, and certification. There are two addenda to this chapter. The first is a copy of the educational signs posted in this project, and the second is a copy of their green cleaning plan.

USGBC LEED-NC INNOVATION AND DESIGN PROCESS, FIVE POSSIBLE POINTS

ID Credit 1–1.4: Innovation in Design (1–4 Points)

The intention of this section is to encourage design teams to exceed the building performance required to fulfill the LEED prerequisites and credits. This credit is not prescriptive; rather, it invites the design team to go beyond what is required, to be creative, and to be innovative.

The Sweetwater Sound design team has gone beyond the LEED requirements in the following areas.

ID Credit 1.1: Innovation in Design—Ice Storage/Reduce Peak Electrical Usage (1 Point)

Sweetwater Sound applied for this credit for the innovative design of the chillers. They are designed so that they run at night during off-peak hours to produce the ice that conditions the air during the day. This design cost 20,500 dollars more in up-front construction costs; however, this investment saves 6,000 dollars per year in reduced electricity expense, making the return on investment completely in just shy of three and one-half years.

ID CREDIT 1.2: INNOVATION IN DESIGN—EDUCATIONAL BUILDING (1 POINT)

Sweetwater Sound applied for this credit for making the building's energy efficiencies and LEED credits transparent in the form of educational materials that are posted throughout the building. They also conduct building tours for interested groups and do a very good job of explaining the efficiencies the building showcases.

See the educational signs that are hanging on the walls at Sweetwater Sound at the end of this chapter.

ID Credit 1.3: Innovation in Design—Water Use Reduction 40 Percent (1 Point)

Sweetwater Sound applied for this credit for taking the Water Efficiency Credits 3.1 and 3.2 10 percent beyond what was required for water use reduction. The extra cost associated with this credit has already been identified in Water Efficiency Credits 3.1 and 3.2.

ID Credit 1.4: Innovation in Design—Heat Island Effect Roof (1 Point)

Sweetwater Sound applied for this credit for taking the Sustainable Sites Credit 7.2: Heat Island Effect: Roof well beyond what was required to satisfy the credit. Sustainable Sites Credit 7.2 requires that 75 percent of the roof surface have a solar reflective index of 78. Sweetwater Sound achieved this index on 100 percent of the roof surface.

Sweetwater Sound could have applied for additional innovation in design credits if they had been available. One is in the **green housekeeping methods** they have adopted for the facility. A copy of their green cleaning plan is included at the end of this chapter. Another innovation in design is in the employee services provided on-site. Some of the services provided are **in-house dining and concierge, workout room and barbershop**. Sweetwater Sound is located several miles from the nearest restaurant, drycleaner, laundry, gym, and so on, so these employee services reduce the amount of driving employees have to do during their working day.

And the final innovation in design is in achieving well over the **open, green space** required by the LEED Sustainable Credit 5.2. The requirement of the credit is to maintain 25 percent, and the open space Sweetwater Sound was able to preserve 43 percent.

Sweetwater Sound Employee Restaurant. Photograph Taken by the Author.

Workout Room for Employees of Sweetwater Sound.   *Sweetwater Sound/John Hopkins. Reproduced by Permission.*

Concierge Service for Sweetwater Sound Employees.   *Sweetwater Sound/John Hopkins. Reproduced by Permission.*

ID Credit 2: LEED Accredited Professional (1 Point)

The intention of this credit is to encourage and facilitate the process of the registration, application, and certification of the project. In order to achieve this credit, there must be a LEED Accredited Professional functioning as part of the design team.

Sweetwater Sound did achieve this credit. Bob Patton, Architect, with MSKTD, an architectural firm headquartered in Fort Wayne, Indiana, was the LEED Accredited Professional on this project.

Summary

Chapter 11 is concerned with the innovation and design LEED credits for which our commercial project has applied. This project has gone beyond the LEED requirements in a number of areas, but four credits are the maximum allowed in this segment. They've applied for credits for reducing peak electrical usage from the grid, for making the building itself educational in the signs that describe the energy efficiencies, for water use reduction that extends well beyond LEED requirements, and for the reduction of heat island effect due to the roof that extends beyond that required to satisfy the earlier credit.

There are a number of other innovations on this project that would have warranted additional credits if they had been available. The final credit in this section is satisfied by the employment of a LEED Accredited Professional to facilitate the process of registration, application, and certification.

There are two addenda to this chapter. The first is a copy of the educational signs posted in this project, and the second is a copy of their green cleaning plan.

Questions and/or Assignments

1. How many possible points can be achieved through Innovation and Design?
2. What is the intention of the Innovation and Design Credits?
3. What are the responsibilities of the LEED Accredited Professional on a project?

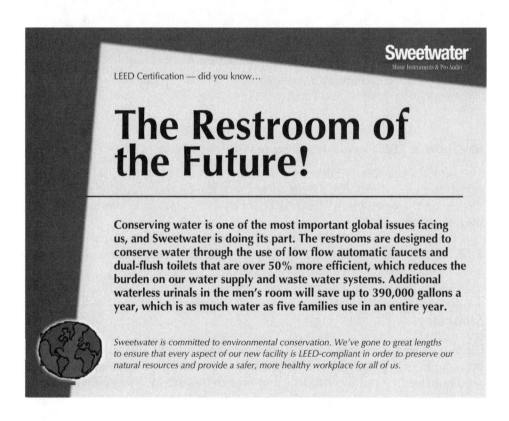

LEED Certification — did you know…

Who turned on the lights?

Have you noticed that when you walk into a dark room the lights come on? There are sensors in almost every room of the building that automatically dim or turn off lights in rooms that are naturally lit or unoccupied, reducing the energy needed to power our building. The amount of natural light in our building creates a relaxing work environment while allowing us to reduce our dependence on electrical lighting.

Sweetwater is committed to environmental conservation. We've gone to great lengths to ensure that every aspect of our new facility is LEED-compliant in order to preserve our natural resources and provide a safer, more healthy workplace for all of us.

Sweetwater
Music Instruments & Pro Audio

LEED Certification — did you know…

Not too hot, not too cold!

We've gone to great lengths using modern technology to make sure that the temperature is just right for you. We installed insulated, fritted glass on the east, south, and west walls to reduce the amount of heat absorbed from the sun, limiting how hard the air conditioning needs to work to keep us cool in the summer. The roof membrane is white to be extremely reflective, cutting the amount of heat absorbed during the summer months.

Sweetwater is committed to environmental conservation. We've gone to great lengths to ensure that every aspect of our new facility is LEED-compliant in order to preserve our natural resources and provide a safer, more healthy workplace for all of us.

Sweetwater
Music Instruments & Pro Audio

LEED Certification — did you know...

Recycle, Recycle, Recycle!

Not only do we recycle everything we can that's generated by our day-to-day business, but we went to great lengths to recycle as much of the waste generated by the construction of our building as possible. From the sorting of scrap metal created during the renovation of the pre-existing structure to the re-use of the old parking lot asphalt as the base of the new lot, we recycled everything we could, and we always will.

Sweetwater is committed to environmental conservation. We've gone to great lengths to ensure that every aspect of our new facility is LEED-compliant in order to preserve our natural resources and provide a safer, more healthy workplace for all of us.

Sweetwater
Music Instruments & Pro Audio

LEED Certification — did you know...

Breath of Fresh Air!

Before we moved in, the building was flushed to remove any harmful vapors and toxins that might have been present due to the construction process. We've also installed air-monitoring systems that will automatically sense a change in air quality, and pump in fresh air from outside when necessary. Our campus is smoke-free, further preserving the air quality. So take a deep breath, and enjoy the fresh air!

Sweetwater is committed to environmental conservation. We've gone to great lengths to ensure that every aspect of our new facility is LEED-compliant in order to preserve our natural resources and provide a safer, more healthy workplace for all of us.

LEED Certification — did you know…

Sweetwater
Music Instruments & Pro Audio

Built to Save Trees!

Even though wood does in fact grow on trees, we went to great lengths to make sure the materials we used didn't have a negative impact on the environment. The wood ceilings found throughout the office areas are composed of a thin wood veneer — harvested in a certified forest — backed with bamboo plywood, a rapidly renewable material. Plus, a minimum of 50% of the wood products used during the construction are certified in accordance with the Forest Stewardship Council's Principles and Criteria.

Sweetwater is committed to environmental conservation. We've gone to great lengths to ensure that every aspect of our new facility is LEED-compliant in order to preserve our natural resources and provide a safer, more healthy workplace for all of us.

Sweetwater
Music Instruments & Pro Audio

LEED Certification — did you know…

Heads Up!

The skylights throughout the building are multi-purpose. Not only do they look cool, but they also allow in an abundance of natural light. The natural light creates a more relaxing environment and reduces our dependence on electrical lighting, saving power.

Sweetwater is committed to environmental conservation. We've gone to great lengths to ensure that every aspect of our new facility is LEED-compliant in order to preserve our natural resources and provide a safer, more healthy workplace for all of us.

Sweetwater
Music Instruments & Pro Audio

LEED Certification — did you know…

Locally Minded

Nearly 25% of the building materials for this facility were obtained within 500 miles of the building. All of the warehouse structural steel and panels were manufactured just north of Fort Wayne; the stone cladding the new recording studio and auditorium spaces is quarried in Glenmont, Ohio. This not only helps support the area economy, but also cuts down on the cost of transporting construction materials to the site, not to mention the use of fossil fuels.

Sweetwater is committed to environmental conservation. We've gone to great lengths to ensure that every aspect of our new facility is LEED-compliant in order to preserve our natural resources and provide a safer, more healthy workplace for all of us.

Sweetwater
Music Instruments & Pro Audio

LEED Certification — did you know…

Rapid Renewal

Much of the wood used in the construction process was selected based on a harvest cycle of 10 years or less. This prevented the construction of our building from negatively impacting the wooded environment.

Sweetwater is committed to environmental conservation. We've gone to great lengths to ensure that every aspect of our new facility is LEED-compliant in order to preserve our natural resources and provide a safer, more healthy workplace for all of us.

Sweetwater
Music Instruments & Pro Audio

LEED Certification — did you know…

Built with Recycling in Mind

We specifically chose the carpeting, fabric, and metals found throughout the building because they were made from recycled materials, reducing the environmental impact of our construction project. We also put an immense amount of effort into sorting and recycling waste generated during the renovation of our building, down to re-using the asphalt from the old parking lot as the base for the new lot!

Sweetwater is committed to environmental conservation. We've gone to great lengths to ensure that every aspect of our new facility is LEED-compliant in order to preserve our natural resources and provide a safer, more healthy workplace for all of us.

Sweetwater Green Cleaning Plan

Purpose

Provide a corporate Standard Operating Procedure (SOP) which ensures consistent, environmentally responsible and sustainable janitorial maintenance services. The intent of this SOP is to minimize exposure of building occupants and maintenance personnel to potentially hazardous chemical, biological, and particle contaminants which may adversely impact air quality, health, building finishes, facility systems, or the environment.

Scope

The scope of this document covers all normal janitorial duties that are undertaken in the course of maintaining the corporate offices.

This scope includes the following:
1) Entryway Systems Maintenance
2) Isolated Chemical Storage and Mixing Areas
3) Sustainable Cleaning Systems including Chemicals and Equipment
4) Use of Concentrates from Dispensing Equipment
5) Carpet Maintenance
6) Disposable Cleaning Products
7) Integrated Pest Control

Requirements
1. Corporate Facility Manager is responsible to provide janitorial maintenance services (internal or contracted) which reduce overall risk and provide a safe and effective work environment, while minimizing environmental impact. The attached guidelines are provided to produce this result in the area of all janitorial services including chemical usage.
2. All operations must meet local regulatory requirements at a minimum.
3. Green janitorial requirements shall be used by all staff who participate in the maintenance of the corporate offices.
4. These green cleaning requirements will apply to all areas of the facility, e.g., private offices, work area, kitchen, training area, warehouse, and lobbies.
5. Results of this program shall be documented by the Corporate Facility Manager and reviewed on an annual basis with the Company President. Annual report shall at a minimum include chemical use listing, safety/incident review, and performance/inspection documents.
6. Standards, product registrations, and housekeeping practices are constantly evolving. The Corporate Facility Manager must keep abreast of developments and strive for continuous improvement in performance and environmental achievement.
7. The Corporate janitorial maintenance policy is defined by the "Green Janitorial Plan" included in this standard.

Contacts and References

U.S. Green Building Council, "LEED for Existing Buildings Reference Guide," 2003.
U.S. Green Building Council, LEED NC Credit Interpretation Ruling dated 4/8/2004.
http://www.usgbc.org

Green Seal's Product Certification standard and list:
Industrial & Institutional Cleaners (GS-37)
http://www.greenseal.org

Green Janitorial Plan

The purpose and intent of the Green Janitorial Plan is to avoid exposure of building occupants and maintenance personnel to potentially hazardous chemical, biological, and particle contaminants which may adversely impact indoor air quality, health, building finishes, and systems, and to minimize the impact of the building maintenance program on the environment. Additionally, this Plan is intended to reduce the risk of both occupants and the company from injury and/or health problems. The promotion of high-quality indoor environment will have positive beneficial effects on employee health and productivity, lifecycle building maintenance costs, and overall environment.

(*continued*)

Entryway Systems Maintenance

Properly installed and maintained entryway systems will greatly reduce the amount of foreign matter tracked into the building, reduce the risk of slips/falls inside the building, and protect the building flooring systems from excessive wear and tear, thereby reducing interior maintenance requirements.

All entryways shall be protected with appropriate mats. At the main entrance, double matting is utilized both inside and outside of the double-door air lock passage way. Mat systems and application shall be specified and applied as seasonally appropriate. For example, in the winter when grit, salt, ice, and water are prevalent, a dual (external/internal) mat system may be required to adequately protect the building, and to supplement the permanent system installed at the main entryway.

The Corporate Facility Manager shall develop specifications and plans for applying, cleaning, and maintaining entryway systems and mats. A log shall be maintained to document that the systems have been effectively maintained. This log and system performance shall be reviewed at least annually by the Facility Manager.

Isolated Chemical Storage and Mixing Areas

Proper isolation, storage, and handling of chemicals will reduce the risk of occupant exposure to potentially hazardous materials.

All cleaning chemicals shall be stored in isolated areas of the building. Proper isolation includes the following:
- Proper ventilation systems to assure direct-to-outside air exhaust, no air recirculation, and negative static pressure in the storage room.
- Hot and cold water supplies and sink drains plumbed for appropriate disposal of liquid wastes.

Only authorized cleaning personnel and the Corporate Facility Manager shall have access to the chemical storage and mixing areas. The Corporate Facility Manager shall maintain building plan drawings indicating all areas where chemical storage and mixing occurs in the building, and shall document appropriate design and maintenance of the supporting building systems. Cleaning practices shall be reviewed annually to insure continued compliance as well as providing opportunities to incorporate improved tasks.

Sustainable Cleaning Systems including Chemicals and Equipment

Janitorial maintenance includes floor care, restroom care, and general cleaning. "Sustainable Cleaning" encompasses more than the concept of minimizing exposure of personnel to potentially hazardous chemicals. To achieve leadership in environmental responsibility within janitorial maintenance systems, the Corporate Facility Manager must consider the life cycle of the building materials and maintenance methods, and incorporate concepts of total cost of performance, safety in use and application, and overall environmental impact. All stages of sustainable building maintenance can be measured for environmental performance, including product selection, installation, operation, long-term maintenance, and eventual disposal.

Environmental and safety aspects of sustainable janitorial maintenance are defined in this plan as follows:

- Facility safety, and health and environmental practices must be compliant with applicable local regulatory requirements.
- The Corporate Facility Manager shall develop and communicate proper disposal methods for all janitorial wastes, including floor care stripping wastes.
- All janitorial personnel shall be properly trained in the use, maintenance and disposal of cleaning chemicals, dispensing equipment, and packaging. Training records certifying each person's specific training dates shall be kept by the Corporate Facility Manager.
- Supplier's Material Safety Data Sheets and Technical Bulletins for all cleaning chemicals shall be provided by suppliers. The suppliers of cleaning products shall provide full disclosure of ingredients on Material Safety Data Sheets. Additionally, suppliers must provide training materials on the hazards and proper use of housekeeping chemicals for workers.
- "Full Disclosure" for products that are not formulated with listed suspect carcinogens is defined as (i) disclosure of all ingredients (both hazardous and non-hazardous) that make up 1% or more of the undiluted product and (ii) use of concentration ranges for each of the disclosed ingredients. "Full Disclosure" for products that are formulated with listed suspect carcinogens is defined as (i) disclosure of all ingredients (both hazardous and non-hazardous) that make up 0.1% or more of the undiluted product and (ii) use of concentration ranges for each of the disclosed ingredients. Suspect carcinogens are those that are listed on authoritative lists available for MSDS preparation: IARC, NTP, and

(continued)

(continued)

California Proposition 65 lists. Concentration range definitions are available from the Canada WHMIS regulation. The intent of the above disclosure requirement is to have a facility disclosure policy that is responsive to the needs of health and safety personnel. If, however, the above disclosure requirement is not met on the MSDS, then disclosure can be provided by suppliers through other means that are easily accessible to health and safety personnel.

- Low environmental impact cleaning products shall be used in accordance with the Green Seal GS-37 standard and/or nationally recognized green certification. In the United States products not covered by GS-37 (such as floor finishes or stripper) shall meet or be less volatile than the California Code of Regulations maximum allowable VOC levels for the appropriate cleaning product category.

- We do not support the use of antibacterial hand soaps as there is now considerable research showing that the use of antibacterial agents contained in soap kill off normal bacteria, creating an environment for resistant, mutated bacteria that can become impervious to antibiotics. No antibacterial hand soap or similar type products will be used in our facility.

- A log shall be kept that details all cleaning chemicals used or stored on the premises (stored products include those that are no longer used, but still in the building). Attachments to the log shall include manufacturer's Material Safety Data Sheets and Technical Bulletins. In locations where Green Seal is a nationally recognized standard, the log shall identify the following:

    - An MSDS and/or label from the manufacturer specifying that the product meets the VOC content level for the appropriate product category as found in the California Code of Regulations.
    - A copy of the Green Seal Certification.
    - If the product has not been certified by Green Seal, the manufacturer will provide test data documenting that the product meets each of the environmental health and safety criteria set forth in Green Seal Standard GS-37.

- When available, chemical concentrates dispensed from closed dilution systems must be used as alternatives to open dilution systems or non-concentrated products.

- Selection of flooring used in the facility, whether a new installation or replacement, shall consider all potential environmental impacts over the full life of the floor system, including raw material extraction and use, installation practices, maintenance requirements, overall useful life, hygiene, appearance and safety attributes, and eventual disposal. A scoring system should be used to develop and evaluate alternatives, including consideration of the total cost of ownership. The selection of flooring materials and their maintenance must consider the full life cycle impacts in order to ensure they will offer the most sustainable floor care system.

- Resilient tile and hard flooring coating systems, including floor finishes and restoration products, shall be slip-resistant (as defined by ASTM Std D-2047) and shall be highly durable in order to maintain an acceptable level of protection and gloss for a minimum of one (1) year before stripping/removal and recoating is necessary.

- A written floor maintenance plan and log will be kept which details the number of coats of floor finish being applied as the base coat and top coats, along with relevant maintenance/restoration practices and the dates of these activities. The duration between stripping and recoat cycles shall be documented.

- A log shall be kept for all powered janitorial equipment. The log should identify the date of purchase and all repair and maintenance activities. Equipment shall meet these requirements:

    - Powered maintenance equipment should be equipped with vacuums, guards, and/or other devices for capturing fine particulates, and shall operate with a sound level less than 70dBA.
    - Propane-powered floor equipment shall have high-efficiency, low-emissions engines.
    - Automated scrubbing machines shall be equipped with variable-speed feed pumps to optimize the use of cleaning fluids.
    - Battery-powered equipment shall be equipped with environmentally preferable gel batteries.
    - Where appropriate, active micro fiber technology shall be used to reduce cleaning chemical consumptions and prolong life of disposable scrubbing pads.
    - Powered equipment shall be ergonomically designed to minimize vibration, noise, and user fatigue.
    - Equipment shall have rubber bumpers to reduce potential damage to building surfaces.

(continued)

Use of Concentrates from Dispensing Equipment

Use of chemical concentrates has several positive environmental benefits:

1. Significantly lower transportation costs between manufacturer and end-user.
2. Significantly lower use of packaging materials.
3. Lower real chemical use to obtain same performance.
4. Potentially lower exposure of maintenance personnel to hazardous chemicals.

Chemical concentrates may present higher hazards upon exposure. Proper containment, storage, and dispensing are critical to avoid employee exposures. Exposure to hazardous chemicals is minimized by using closed dispensing systems. Concentrates sold for manual dilution in buckets or bottles can actually increase the risk of employee exposure.

Chemical concentrates dispensed from closed dilution systems shall be used preferentially to open dilution systems or non-concentrated products. Chemicals and their use shall comply with all directives shown in Section 3.

If equipment is used to control the dilution of concentrated cleaning chemicals, then a log shall be kept which includes the equipment manufacturer's technical information, as well as the date of installation, maintenance, and repairs. The log shall also contain the desired dilution rates for each cleaning product and a plan for maintaining the desired dilutions on an annual basis.

Janitorial personnel shall be properly trained in the use, maintenance, and disposal of cleaning chemicals, dispensing equipment, and packaging.

Carpet Maintenance

Low environmental impact janitorial equipment includes the use of durable carpet care equipment, such as upright, backpack, and wide area vacuums equipped with power-heads and capable of capturing 99% of particulates 0.3 microns in size. Carpet extraction equipment shall be capable of removing sufficient moisture such that carpets can dry in less than 24 hours. Carpet care equipment shall be electric or battery powered and shall have a maximum sound level less than 70dBA.

Wherever possible, a dry encapsulation method shall be used to reduce chemical use and drying time.

A log will be kept which details the relevant maintenance/restoration practices and the dates of these activities. The duration between extraction cycles shall be documented.

A log will be maintained which lists all carpet care equipment including vacuums (e.g., upright, backpack, wide area, and wet/dry) and equipment used for maintaining resilient and hard floors (e.g., buffers, burnishers, and auto-scrubbers). Documentation will be kept on each piece of equipment identifying performance capabilities.

Disposable Cleaning Products

Low environmental impact janitorial supplies will include the use of disposable paper (toilet tissue and paper towels) utilizing 100% recycled content and a minimum of 30% postconsumer recycled content, AND which are manufactured without the additional use of elemental chlorine or chlorine compounds (Processed Chlorine Free). Plastic trash can and other liners will utilize a minimum of 10% postconsumer recycled content.

Purchasing records manufacturer's technical bulletins for paper and plastic lines, which indicates grade, total recycled content, postconsumer recycled content, and bleaching processes (if applicable) shall be provided.

Low-Impact Integrated Pest Control

Low environmental impact integrated pest management shall consist of a written pest management plan that details the techniques, strategies, and schedules to control unwanted pests. The plan will include a log of all pest management activities, including but not limited to establishing barriers, setting of traps, use of baits, and chemical applications. Individual product Material Safety Data Sheets and Technical Bulletins will be maintained on all pest control products and be attached to the plan.

12

■ ■ ■

Sustainable Construction: A Collaborative Project

Abstract

Chapter 12 is a description of a sustainable residential construction project for Habitat for Humanity. This chapter describes the process collaborators went through in order to make this home sustainable. It covers the special needs of the Habitat family, the design charrette, the site, and the design process. There is an addendum containing a full set of blueprints for the house at the end of the chapter.

For many years, Habitat for Humanity has wanted to change the way they build homes to make them sustainable. The Executive Director of Fort Wayne's Habitat for Humanity, Don Cross, expressed how difficult this change would be because they are volunteer based. It is a massive endeavor to think about having to train and retrain an entire volunteer force the size of Habitat's. This project addresses that problem, not just here in Fort Wayne, Indiana, but throughout the United States. Seniors enrolled in the Construction Engineering Technology, Bachelor's degree program, through the Center for the Built Environment (CBE) at Indiana University Purdue University Fort Wayne (IPFW; http://www.ipfw.edu), are collaborating with members of the Northeast Indiana Green Build Coalition (NEIGBC; http://www.neigbc.org), and our local Habitat for Humanity chapter (http://www.fortwaynehabitat.org), to design and build a sustainable residence.

As we've mentioned already, one of the ways in which sustainable construction is different from conventional building is that it depends on so much front-end collaborative, integrated design work. During the earliest stage, the focus is on building relationships between the people who represent the different competencies, and on working together to achieve the design goals for energy efficiencies. Because we would like to do something to help educate the populace about this sustainable design and construction process, we are filming the process and are producing an educational DVD for use by Habitat for Humanity groups throughout the United States.

Sustainable homes have been designed and built for Habitat families in other locales, so we aren't the first ones to do this. But the training would take place and the build would happen, and then when it came time to build the next home, so many staff members and volunteers were different that it was as if no training in sustainable construction had happened at all. Our intention is that as turnover occurs in Habitat for Humanity volunteers, board members, and staff, retraining in the sustainable building process will be made easy by using the DVD that is being produced here. In this way, we will be able to assist Habitat for Humanity in their desire to ensure that every home they build is sustainable.

Let me just briefly explain how the Habitat program works. As I've talked with groups about this project, I have learned that many people don't actually understand it. This is a very good program that works well and has a high success rate. Habitat helps families to buy and own a house, who ordinarily wouldn't be able to do so. There are a number of different parameters Habitat uses to discern whether a family qualifies for the program or not. Items like income level and family size, or when several generations of family are living together in one house are considered. Once accepted into the program, those who qualify usually need to work with a credit advisor to help them work through the process of cleaning up their finances. That can often take up to a year. Once their credit is cleared up, they attend workshops that instruct them in home ownership and maintenance, and each adult in the family begins working on the 400 volunteer hours they must put in to develop their sweat equity. That is what serves as the down payment. Their volunteer hours can be served working on Habitat homes that are currently being built, or if they aren't physically able to swing a hammer or sling paint, they can work on projects in the office like getting mailings together. It's a great program that serves its mission well.

The NEIGBC began working on this collaboration by conducting a brain-storming session to discuss the ways in which the group would be able to contribute to the project. A variety of coalition members, experts in different sustainable competencies, volunteered to come to the classroom and mentor students about their individual areas of expertise. One mentor talked with the students about site energy mapping. Another mentor talked about how to orient the home on the site so that summer shading and winter sun could be captured and so that summer breezes could be invited and winter wind blocked. A third mentor talked about capturing the sun's energy using photovoltaic panels and a solar hot water heater. We also had mentors make presentations to the class about water resources, building envelope (walls, roof, and windows), insulated concrete forms (ICFs), and sustainable landscaping. The Executive Director of Habitat for Humanity, their Project Manager, and one of the board members also came to the class to explain how the Habitat program works, in terms of their normal construction process.[1]

We invited the Habitat family to come to class as well. This family came from Burma originally, and they are here in the United States as political refugees. The students were able to ask the family questions about their lives in Burma, what their home had been like, and what they enjoyed doing. The woman spoke about how lucky they had been in terms of the home they had lived in because it had a metal roof, and they were able to have this "expensive house" because her family had inherited it from an uncle. She also told us about the day her father had been called in to the central political office for what he was told would be a mediation session. When he arrived he was shot in the head, and she never saw him again. She told us that she immediately joined the army, and from then on she lived in the jungle and her house was made primarily of bamboo. She described to us how the roof was made of large leaves that were woven together, and then layered row by row on top of each other. The whole house could be quickly assembled and just as quickly taken apart when they moved from one area to another.

Outdoors, they expressed a desire for a space to grow their own vegetable garden. They talked about needing a door that would close the kitchen off from the rest of the house because (she said, with a grin) though their food tastes delicious, it smells stinky while cooking! They also talked about their preference for hard surface floors because they're easier to keep clean. They need one of the bedrooms removed as far from the living space as possible because her husband works nights and sleeps during the day. The woman's life is extremely active— she works part time for the city school system as a translator for the teachers and Burmese parents and children. Much of the remainder of her day is spent on the phone. She said that she often has a home phone in one hand and her cell phone in the other because she works as the unofficial volunteer translator for the Burmese immigrant population as well. She is often called to go to doctor's appointments with people or must rush to the emergency room to help people communicate. Her days are filled with service to her fellow immigrants, all of whom are here as political refugees.

At the design charrette that we conducted for this home, this Burmese woman told us that there had been a number of families who were trying to get all of their Habitat volunteer hours and workshops completed in order to be the family who would get to buy "the green house." She had tears in her eyes when she told us that she was so thankful that she and her family were the ones who would be able to buy this house. She said "we came from a green house made of bamboo and leaves, and we get to move into a green house here." We conducted a design charrette, formed of participants from all three collaborating groups. Our students and faculty members from the CBE at IPFW, the family that will be purchasing the home through Habitat, and others from the Habitat staff, board, and volunteers, together with a diverse group of NEIGBC members representing all competencies of the building process, worked together to design the most sustainable residence possible.

Photograph Taken During a Break in the Design Charrette.   *Elmer Denman, Photographer. Reproduced by Permission.*

The design charrette produced several good options. Students took those designs and began working with them to find a design that incorporated all of the natural energies available on site and used as many of the energy efficiencies from the charrette options as possible. See the site design the students produced at the end of this chapter. The ground slopes to the southwest. There are mature trees that are approximately 30 feet tall on the south side of the lot. There are also a couple of mature trees and some shrubbery on the northwest side. Rainwater that falls on this land is channeled to a rain garden that we've built in the low area at the back of the lot on the southwest side. You can find the rain garden design, with plantings identified at the end of this chapter. The rain garden is sized so that all of the water that falls on this land will stay there and be handled by the rain garden. None of it will move into the city storm sewer. (In a city that has been experiencing "100 year floods" every couple of years, this is important!)

The street is on the north side of the lot, and orienting the living areas of this home to take maximum advantage of the passive energies available is one of the most important features. The typical layout of a home with the front door and living room facing the street was rejected. Instead, to take advantage of the passive warmth from the sun in the winter, the living area is situated on the south side of the house, with the main entrance placed on the east. Much of the south side of the house is glazed to capture the winter sun, and will be shaded by the trees during the summer.

The students investigated using ICFs to build the 4 feet-high crawl space. But ultimately the Habitat staff and Board rejected that idea because they felt that using ICFs would compromise volunteer involvement, an important

Newly Planted Rain Garden, Student, Mary Kopke, Designer, and Mark Ringenberg, Hoosier Releaf, Donor of All Plantings and Materials.

aspect of their homebuilding process. Instead of ICFs, we used concrete block with enough interior insulation to ensure an *R*-value of 40. Because the ground slopes so much on this lot, we were able to install an exterior entrance door to the crawl space on the southwest side of the structure.

The walls of the house have been built using 2 × 4 staggered-stud wood framing, and walls and ceiling/attic are insulated to an *R*-value of 40. All upright 2 × 4 members are also finger-joined. The exterior walls are constructed with an outside face built with 2 × 4's, 24 inches on center, while the inside face is built with 2 × 4's, 16 inches on center. There is a half-inch space between the outside and inside facing studs. Building in this way prevents thermal breaks from occurring in the wall. (See photograph below.) In the normal construction of a wood-framed house, 2 × 4 or 2 × 6 studs are spaced 16 inches on center and insulation of some variety is put in between the studs. But in this design, every stud is an un-insulated place that allows heat transfer to occur. In the student's design, every exterior wall is fully insulated and thermal breaks are reduced to a minimum.

Daylighting is provided for this home in a couple of ways. At the peak of the living room is a three-window clerestory that faces east, and there are two solar tubes—one in the bathroom and another in the hallway, from which the bedroom doors open. These features mean that it is unnecessary to turn lights on most days until the evening hours. Another feature of the clerestory

Staggered Stud Exterior Wall, Finger Joined Studs.   *Dale Miller, Photographer.*

windows is that they are operable with a hand crank, which creates a nice exhaust of heat and convection cooling during warmer weather.

The HVAC system that augments the passive heating and cooling is WaterFurnace's Envision, a 500 percent efficient geothermal unit. They donated this furnace to the project, along with all installation to the exterior of the structure. Two vertical loops were installed because space is at a premium on this small lot. In the process of fitting the unit to the house, they realized that the smallest unit they make is sized to heat and cool a home that is approximately twice the size of this Habitat house. Because of that, they are investigating building some smaller units. You can gain more information and specifications about the geothermal Envision by visiting the WaterFurnace website at http://www.waterfurnace.com.

Another local Company, Home Guard Doors and Windows, made the energy-efficient windows for the home. Normally, the *R*-value of windows ranges from 0.9 to 3.0, and *R*-values are a measure of heat flow resistance. The lower the *R*-value on the window, the more heat will be lost though it during the winter, and the more heat will be gained during the summer—exactly the opposite of the desired conditions. The windows that were installed in this home have an *R*-value of 10. They have an invisible, metallic coating that blocks 70–75 percent of radiant heat from escaping in the winter, by reflecting it back into the room, and 25 percent of the radiant heat is

Clerestory Above Living Room. *Dale Miller, Photographer.*

reflected back to the outside during the summer. This metallic coating is called low emissivity (low-E).

The countertops are concrete and were made by the students. First we made the bathroom countertop in the Construction Lab at IPFW. Students built the form for the counter using 0.75-inch white melamine, and created a knockout for the sink with 2-inch blue foam covered with clear plastic kitchen wrap. They used black silicone caulk on all seams, sprayed the bottom of the form with tack spray, and sprinkled some colorful aggregate onto the tack. They reinforced the concrete by suspending rebar and fencing material 1 inch from the bottom of the form. Students mixed the concrete and color, poured it into the form, and vibrated the concrete to eliminate all air spaces. Four days later, we removed the form and sink knockout, and turned it over to reveal the top side. The students used a grinder on the surface until the aggregate showed through, used slurry on some imperfections, and after that was dry, they used a grinder on the surface again. When they were satisfied with the surface, they used a sealer, and finished it with beeswax and buffing.

The students also made the concrete countertop for the kitchen, but this time they did it by casting it in place. The students wanted to put linoleum flooring in this house, but it would have taken the project over budget, so instead, the installed flooring is tile made with recycled content, in kitchen, bathroom, and

Students Working on the Precast Bathroom CounterTop in the Lab at IPFW.   *Elmer Denman, Photographer.*

utility room. The living room, hall, and bedrooms are floored with carpeting made with 25 percent recycled content.

Cabinetry in the kitchen and bathroom is made of rubberwood, a reclaimed hardwood that is a byproduct left when all of the latex has been removed from the wood. Despite the misconceptions the name might invoke, rubberwood is one of the most durable of hardwood lumbers. It is a member of the maple family and has a very dense grain, and very little shrinkage when being dried. It is also an ecologically sustainable lumber because after latex has been harvested for 26–30 years, the trees are usually felled and new trees are planted. The felled trees used to be burned; now they are harvested and used primarily to make furniture and cabinets.

All interior materials are of either no or low volatile organic compound (VOC). The paint is from Sherwin Williams Harmony line, which is a low-odor, no VOC formula, containing fewer solvents. Sherwin Williams uses sustainable raw materials in their paints, like soy and sunflower oils, and have significantly reduced the amount of solvent in their paint so that fewer vapors are emitted into the atmosphere.

This project has not only been extremely instructive for students, but has also made a deep contribution to our society by furthering our knowledge of sustainable design and construction, and also by helping Habitat for Humanity build energy-efficient homes. The DVD of the process of this project was sent out to Habitat locations throughout the United States in May 2008.

Rubberwood Kitchen Cabinets Being Installed by a Habitat Volunteer; Photograph Also Shows Student-Made Kitchen Concrete Counter.

Summary

This chapter is about a sustainable residential construction project for Habitat for Humanity. This chapter describes the process collaborators went through in order to make this home sustainable. It covers the special needs of the Habitat family, the design charrette, the site, and the design process. The chapter also describes the particulars in both design and construction that make this home energy efficient. Some of the important features of the home are the orientation enabling passive energy use, staggered stud exterior wall, insulated, conditioned-air crawl space, elimination of thermal breaks in exterior walls, operable clerestory windows that allow the possibility of convection cooling, a WaterFurnace Envision geothermal unit, rubberwood cabinetry, and student-made concrete countertops. This chapter has an addendum containing a full set of blueprints for the house.

Questions and/or Assignments

1. Investigate your own home, regardless of where it is that you live (apartment, dorm room, condominium, etc.), and determine the insulative R-value. Then figure out what you could do to increase the R-value to at least 40.

2. As an alternate or addition to assignment 1, do the same investigation on one of your university's buildings.
3. Conduct the same investigation on the roof of your home and then on the roof of the school's building. What could be done to increase the *R*-value to 40?
4. What are thermal bridges? Give an example of some ways to eliminate them when you are building.
5. What is convection cooling? How can you construct a building so that convection cooling occurs?

Notes

1. The participating mentors were Nate Rumschlag, PE, Steve Park, Architect, Matt Kubik, Architect and IPFW Faculty, Candace Imbody, Owner of Construction Recycling Solutions, Doug Ahlfeld, solar hot water and photovoltaic expert, Dan Ernst, Owner of EarthSource, Inc. and Heartland Restoration, Jerry Yoder, ICF Builder, Don Cross, Executive Director of Habitat for Humanity, Lynn Weaver, Project Manager, and Dan Rickert, a member of the Board for our local Habitat for Humanity.

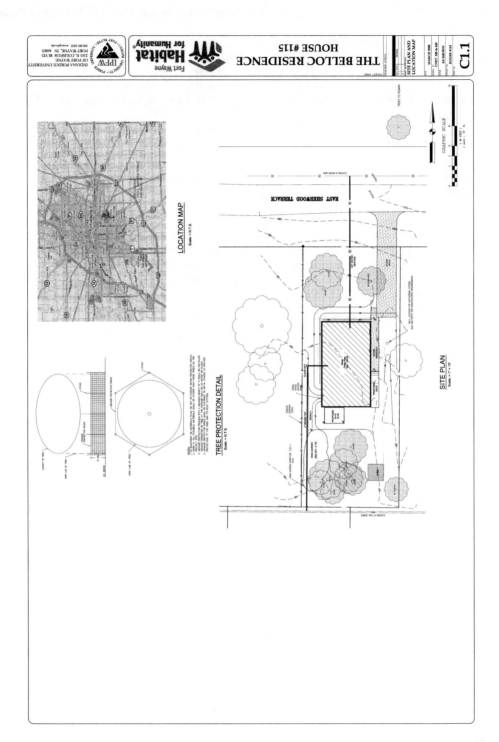

THE BELLOC RESIDENCE
HOUSE #115

SITE PLAN AND
LOCATION MAP

C1.1

MARCH 2008
CNET: 345 & 445
AS SHOWN
HOUSE #115

Habitat for Humanity®
Fort Wayne

INDIANA PURDUE UNIVERSITY
OF FORT WAYNE
2101 E. COLISEUM BLVD
FORT WAYNE, IN 46805
260.481.5832   www.ipfw.edu
UNIVERSITY • FORT WAYNE
IPFW

LOCATION MAP
Scale = N.T.S.

TREE PROTECTION DETAIL
Scale = N.T.S.

SITE PLAN
Scale = 1" = 10'

GRAPHIC SCALE

EAST SHERWOOD TERRACE

TREES TO REMAIN

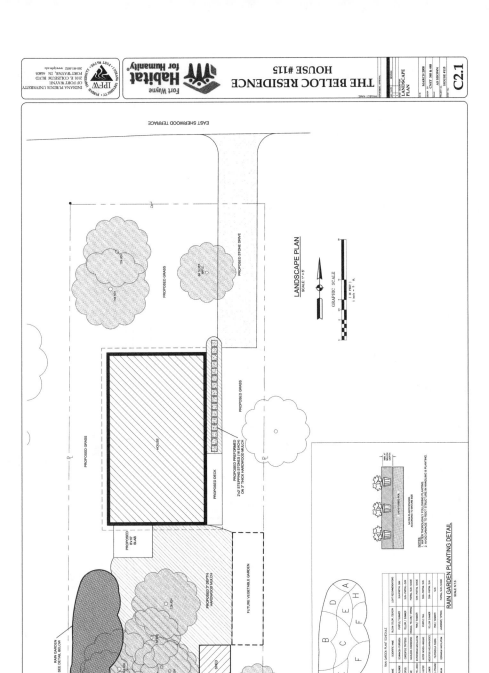

Fort Wayne
Habitat for Humanity®

IPFW
INDIANA-PURDUE UNIVERSITY
OF FORT WAYNE
2101 E. COLISEUM BLVD
FORT WAYNE, IN 46805
260.481.6472 www.ipfw.edu
IPFW • PORT WAYNE • INDIANA-PURDUE UNIVERSITY

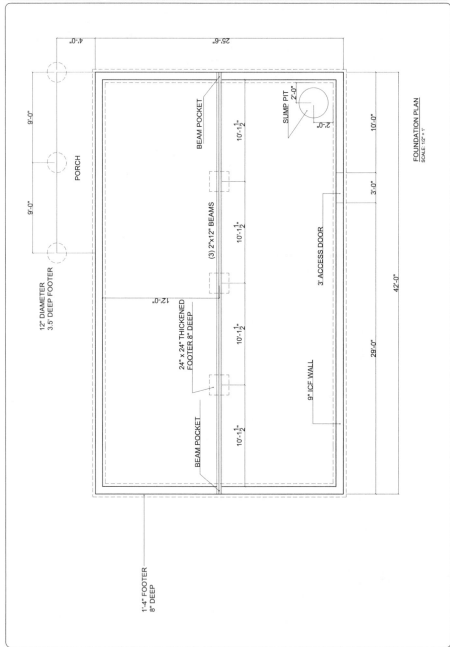

FOUNDATION PLAN
SCALE 1/2" = 1'

C3.1
FOUNDATION PLAN
MARCH 2008
UNIT 360 & 480
AS SHOWN
HOUSE #115

4'-0"
25'-6"
9'-0"
PORCH
9'-0"
12" DIAMETER
3.5" DEEP FOOTER
BEAM POCKET
SUMP PIT
2'-0"
2'-0"
10'-1½"
10'-1½"
(3) 2"x12" BEAMS
10'-1½"
10'-0"
3'-0"
3' ACCESS DOOR
12'-0"
24" x 24" THICKENED
FOOTER 8" DEEP
10'-1½"
42'-0"
29'-0"
9" ICF WALL
BEAM POCKET
10'-1½"
1'-4" FOOTER
8" DEEP

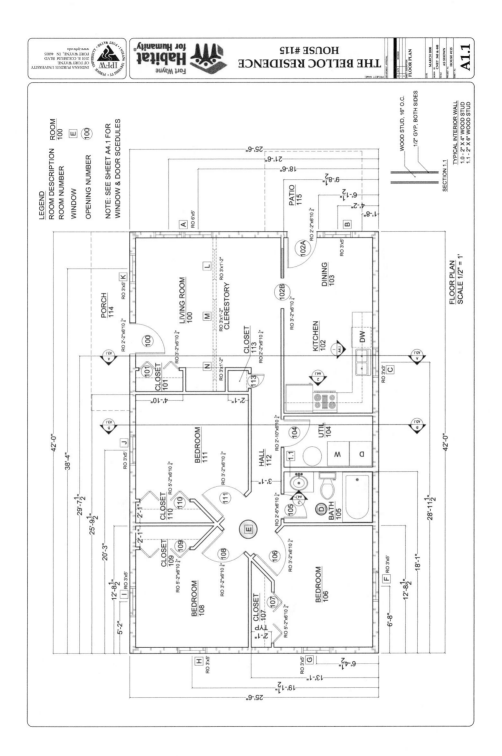

THE BELLOC RESIDENCE
HOUSE # 115

A1.1

MARCH 2006
CNET 340 & 640
AS SHOWN
FLOOR PLAN
HOUSE #115

Fort Wayne
Habitat for Humanity®

INDIANA PURDUE UNIVERSITY
OF FORT WAYNE
2101 E. COLISEUM BLVD
FORT WAYNE, IN 46805
www.ipfw.edu

LEGEND
ROOM DESCRIPTION | ROOM | 100
ROOM NUMBER
WINDOW | E
OPENING NUMBER | 100

NOTE: SEE SHEET A4.1 FOR
WINDOW & DOOR SCEDULES

TYPICAL INTERIOR WALL
1.0 - 2 X 4" WOOD STUD
1.1 - 2 X 6" WOOD STUD

WOOD STUD, 16" O.C.
1/2" GYP, BOTH SIDES
SECTION 1.1

FLOOR PLAN
SCALE 1/2" = 1'

PORCH 114
LIVING ROOM 100
CLERESTORY
CLOSET 113
CLOSET 101
BEDROOM 111
CLOSET 110
CLOSET 109
BEDROOM 108
HALL 112
CLOSET 107
BEDROOM 106
BATH 105
UTIL 104
KITCHEN 102
DINING 103
PATIO 115

139

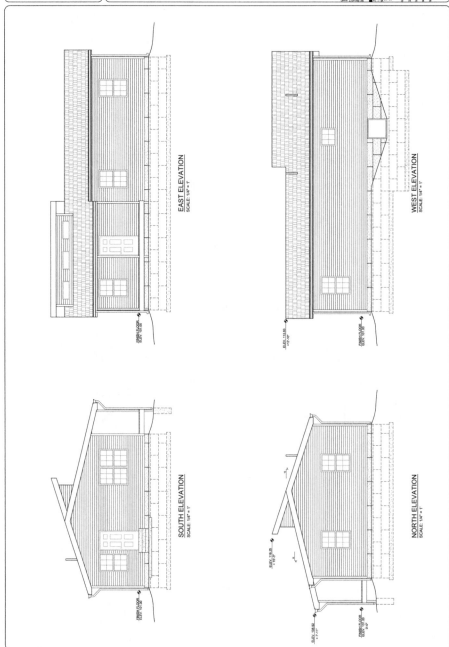

Fort Wayne
Habitat for Humanity®

THE BELLOC RESIDENCE
HOUSE # 115

BUILDING
ELEVATIONS

MARCH 2008

A2.1

INDIANA PURDUE UNIVERSITY
OF FORT WAYNE
2101 E. COLISEUM BLVD.
FORT WAYNE, IN 46805
260.481.6822 www.ipfw.edu

EAST ELEVATION
SCALE: 1/4" = 1'

WEST ELEVATION
SCALE: 1/4" = 1'

SOUTH ELEVATION
SCALE: 1/4" = 1'

NORTH ELEVATION
SCALE: 1/4" = 1'

THE BELLOC RESIDENCE
HOUSE #115

BUILDING
SECTIONS

A3.1

MARCH 2009
UNIT 340 & 440
AS SHOWN
HOUSE #115

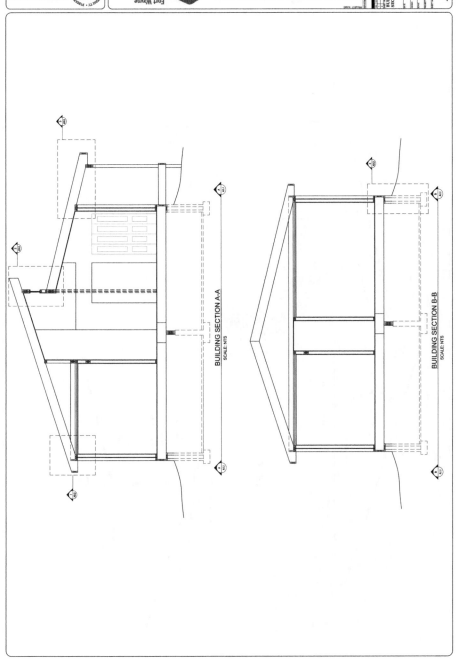

BUILDING SECTION A-A
SCALE: NTS

BUILDING SECTION B-B
SCALE: NTS

A3.2

BUILDING
SECTION
DETAILS

MARCH 2000
CNBT 348 & 448
AS SHOWN
HOUSE #115

THE BELLOC RESIDENCE
HOUSE #115

Fort Wayne
Habitat for Humanity®

INDIANA PURDUE UNIVERSITY
OF FORT WAYNE
2101 E. COLISEUM BLVD
FORT WAYNE, IN 46805
260.481.6821   www.ipfw.edu

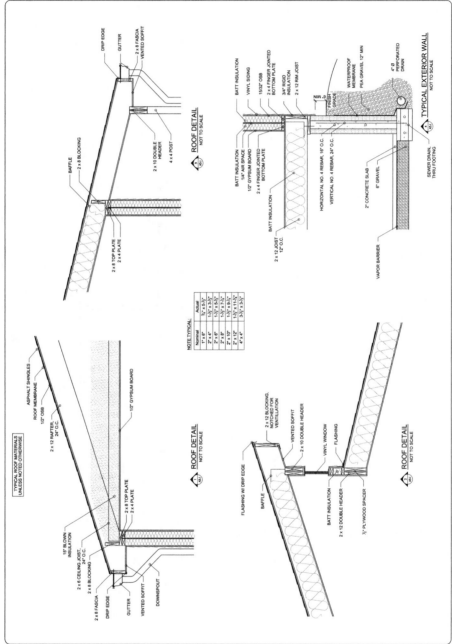

Nominal	Actual
1" x 6"	3/4" x 5-1/2"
2" x 4"	1-1/2" x 3-1/2"
2" x 6"	1-1/2" x 5-1/2"
2" x 8"	1-1/2" x 7-1/2"
2" x 10"	1-1/2" x 9-1/2"
2" x 12"	1-1/2" x 11-1/2"
4" x 4"	3-1/2" x 3-1/2"

NOTE TYPICAL

ROOF DETAIL
NOT TO SCALE

TYPICAL EXTERIOR WALL
NOT TO SCALE

TYPICAL ROOF MATERIALS
UNLESS NOTED OTHERWISE

THE BELLOC RESIDENCE
HOUSE # 115

DOOR/WINDOW/
FINISH ROOM
SCHEDULES &
DETAILS

A4.1

MARCH 2008
CNST 340 & 448
AS SHOWN
HOUSE #115

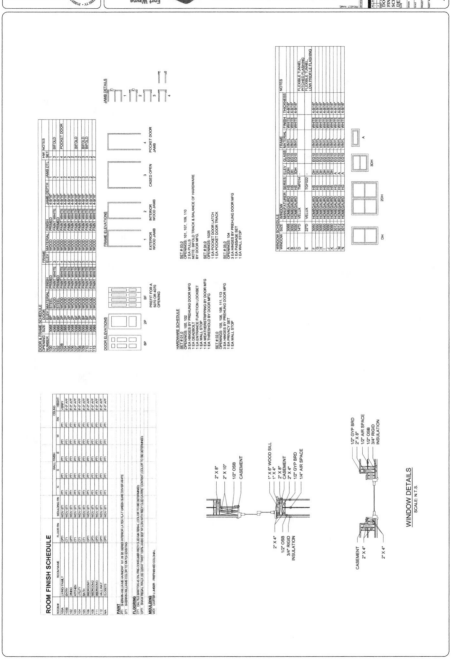

ROOM FINISH SCHEDULE

DOOR & FRAME SCHEDULE

FRAME ELEVATIONS

DOOR ELEVATIONS

JAMB DETAILS

HARDWARE SCHEDULE

WINDOW SCHEDULE

WINDOW DETAILS
SCALE: N.T.S.

THE BELLOC RESIDENCE
HOUSE # 115

Habitat
for Humanity
Fort Wayne

E1.1

INDIANA PURDUE UNIVERSITY
OF FORT WAYNE
2101 E. COLISEUM BLVD
FORT WAYNE, IN 46805
www.ipfw.edu

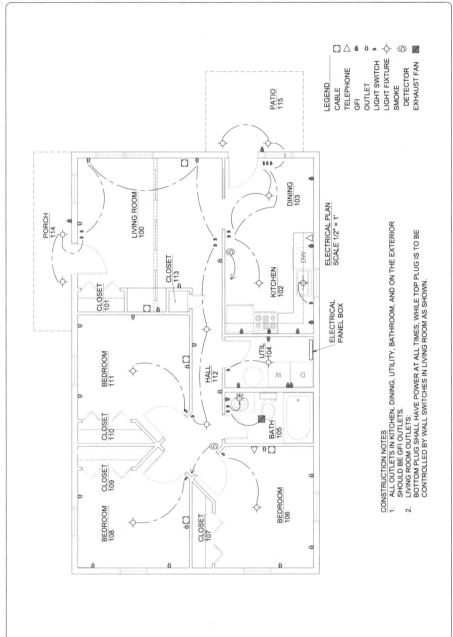

LEGEND
CABLE
TELEPHONE
GFI
OUTLET
LIGHT SWITCH
LIGHT FIXTURE
SMOKE DETECTOR
EXHAUST FAN

PATIO
115

DINING
103

PORCH
114

LIVING ROOM
100

CLOSET
113

CLOSET
101

KITCHEN
102

ELECTRICAL PLAN
SCALE 1/2" = 1'

ELECTRICAL
PANEL BOX

BEDROOM
111

HALL
112

UTIL
104

BATH
105

CLOSET
110

CLOSET
109

CLOSET
107

BEDROOM
108

BEDROOM
106

CONSTRUCTION NOTES
1.   ALL OUTLETS IN KITCHEN, DINING, UTILITY, BATHROOM, AND ON THE EXTERIOR
      SHOULD BE GFI OUTLETS.
2.   LIVING ROOM OUTLETS:
      BOTTOM PLUG SHALL HAVE POWER AT ALL TIMES, WHILE TOP PLUG IS TO BE
      CONTROLLED BY WALL SWITCHES IN LIVING ROOM AS SHOWN.

144

13

■ ■ ■

Sustainable Construction Template

Design Concern	Sustainable Building Approach
Building Envelope	**1.** Choose structure and enclosures that minimize energy loss and maximize thermal barrier 　　Maximize thermal barrier—how? 　　Eliminate thermal bridges—how?
Site	**2.** Choose an energy-efficient location 　　Infrastructure present 　　Mass transportation easily accessible
	3. Inventory existing vegetation 　　Protect vegetation natural to this ecosystem 　　Eliminate invasive species
	4. Place building with least impact possible to site
	5. Any trees that must be removed: 　　Either recycle to a local mill 　　Or chip and use mulch on site
	6. Recycle all construction "waste" 　　Provide bins for different materials scraps 　　Provide recycling barrels for personal cans/bottles
	7. Landscape with native plants and no mow lawns
Building Oriented on Site	**8.** Capture the natural energy: 　　Passive solar heat 　　　　Promote heat gain in cold weather 　　　　Prevent heat gain in warm weather 　　Breezes 　　　　Prevent cold winds in winter 　　　　Promote cool breezes in summer

(continued)

(continued)

Design Concern	Sustainable Building Approach

	Views and daylight
	    Capture pleasing views
	    Capture as much daylight as possible
	Block noise pollution
	Major roof areas face south
	    Plan for photovoltaic Panels and/or solar hot water

HVAC Systems	**9.** Use systems that heat and cool mass rather than air Geothermal or radiant, floor or baseboard, for example
	10. Energy Recovery Ventilation System

Energy	**11.** Design to reduce loads
	    Eg., Low-voltage lighting + daylighting +
	    lighting sensors
	Solar power with battery storage
	Wind generated power with battery storage
	Intertie systems—share and utilize energy with/from grid

Water: Potable, Gray, Black	**12.** Design to use potable water for necessary uses only

	13. Design gray water systems for all other uses
	    Rainwater cachement systems
	    Captured gray water from all allowable sources

	14. Treat waste water on site
	    Create biological filtering systems
	    Move cleaned water back to the water table

Future of the Planet	**15.** Build with durability as the goal (as if you were building a cathedral)

	16. Design the next life into the first one
	    Design for disassembly and reuse, or
	    Use biodegradable materials
	        Wood, concrete, etc.

	17. Use biological filters
	    Trees to clean the air
	    Bioswales to clean residues from vehicles
	    Native vegetation to add nutrients to the soil

	18. Use only nontoxic materials No man-made substances such as PCBs, CFCs, halons, etc.
	19. Reduce consumption of materials Use recycled content materials
	20. Replant trees equal to the amount of lumber consumed
	21. Use natural resources no faster than their own rate of regeneration

I would like to thank Carol Tunnel, Architect, for providing the original inspiration for this checklist.

INDEX

Accredited Professional
  innovation and design, 115
  LEED® certification, 115
Adhesives
  indoor quality, 100
  low-emitting materials, 100
Agrifiber products
  indoor quality, 102
  low-emitting materials, 102
Air. *See also* Ventilation
  indoor quality, 97–103
Alternative transportation, LEED® certification, 52–54
Approaches, sustainable construction, 145–146
Architecture
  communication, 26–29
  inspiring spaces, 29
  interconnectedness, 26–29
  whole systems thinking, 25–30
Area surrounding site, site research, 47
ASHRAE Advanced Energy Design Guide,
    optimizing energy performance, 80–82
ASHRAE standards
  indoor quality, 97–99
  windows, 97
Aspiration, collaboration, 33–34
Automated windows, 77

Bicycle storage, LEED® certification, 53
Biomes
  environmental issues, 43–46
  interdependency of organisms, 43–46
  nitrogen cycle, 45
  nutrient chain, 45
  sites, 43–46
  water cycle, 45
'bodymind', 11
Brownfield redevelopment, LEED® certification, 52
Buckminster Fuller, R., 3
Building envelope, sustainable construction
    template, 145
Building orientation, 76
  sustainable construction template, 145–146
Building reuse, resources, 89–90

Cabinetry material, rubberwood, 133–134
Carpet systems
  indoor quality, 101
  low-emitting materials, 101
Certification process, LEED® certification, 48–49
Certified wood, resources, 94–95
Changing rooms
  LEED® certification, 53, 54
  transportation, 53, 54
Charrettes, design, 128–129
  integrated design, 28–29, 31–38
  outcomes, 31–38
Chemicals/pollutants, indoor quality, 102–103

Chillers, innovation and design, 112–113
Cleaning, Green Cleaning Plan, 122–125
Coatings/paints
  indoor quality, 100–101
  low-emitting materials, 100–101
Collaboration
  aspiration, 33–34
  dialogue, 35
  learning organizations, 33–35
  mental models, 34–35
  personal attributes, 33–37
  personal mastery, 34
  reflective conversation, 35
  sustainability in action, 31–39
Collaborative project, sustainable construction,
    126–146. *See also* Habitat for Humanity
Collection, recyclables, 88
Commissioning, enhanced, energy systems, 83
Communication
  architecture, 26–29
  interconnectedness, 26–29
  LEED® certification, 116–121
  Sweetwater Sound, 116–121
  whole systems thinking, 25–30
Community connectivity, LEED® certification, 52
Composite wood
  indoor quality, 102
  low-emitting materials, 102
Conscious representatives, nature's, 9–14
Construction activity pollution prevention, LEED®
    certification, 50
Construction Indoor Air Quality (IAQ) Management
    Plan, 99–100
Construction waste management, 90–91
Conventional project organization, unconnectedness,
    25–29
Cooling, passive
  buildings, 76–78
  ERVs, 77
  Sweetwater Sound, 76–77
Cost, Lloyd's Crossing, 19

Data collection, site research, 46–48
Daylight views, indoor quality, 104–106, 130–132
Degenerative design, vs generative design, 11–12
Density of development, LEED® certification, 52
Design and innovation. *See* Innovation and design
Development density, LEED® certification, 52
Dialogue, collaboration, 35
Dioum, Baba, 9
Dymaxion House, reusable house, 22–23

Ecology, living systems, 6–7
Ecosystems
  environmental issues, 43–46
  interdependency of organisms, 43–46
  nitrogen cycle, 45

Ecosystems (*continued*)
 nutrient chain, 45
 sites, 43–46
 water cycle, 45
Effluent, water resources, 71
Electricity usage
 innovation and design, 112–113
 reducing, 112–113
Embodied energy, materials, 87–88
Energy mapping/modeling
 site research, 47–48
 software, 48
 solar energy, 48
 wind energy, 48
'energy means'
 cost, 19
 living within, 19–20
 Lloyd's Crossing, 19–20
Energy prerequisites, HVAC systems, 76–86
Energy recovery ventilators (ERVs), passive cooling, 77
Energy, renewable. *See* Renewable energy
Energy systems. *See also* Renewable energy
 enhanced commissioning, 83
 fundamental commissioning, 78
 geothermal unit, 131
 green power, 85
 Habitat for Humanity, 128–129, 131
 LEED® certification, 78–85
 measurement & verification, 85
 minimum energy performance, 79
 optimizing energy performance, 80–82
 refrigerant management, 79, 83–84
 on-site renewable energy, 82
 sustainable construction template, 146
 Sweetwater Sound, 76–86
 WaterFurnace Envision, 131
Enhanced commissioning, energy systems, 83
Environmental issues
 biomes, 43–46
 ecosystems, 43–46
 nitrogen cycle, 45
 nutrient chain, 45
 sites, 43–46
 water cycle, 45
Environmental tobacco smoke (ETS), indoor quality, 97–98
Envision (WaterFurnace's), geothermal unit, 131
EQUEST software, optimizing energy performance, 80–82
ERVs. *See* Energy recovery ventilators
ETS. *See* Environmental tobacco smoke

Floodplain search, site research, 46–47
Fuel-efficient vehicles
 LEED® certification, 53
 transportation, 53
Future of the planet, sustainable construction template, 146

Garbage, 21–22
Generative design
 Lloyd's Crossing, 15–21
 vs degenerative design, 11–12, 15–24

Geodesic dome, 3–4
Geothermal unit
 Merry Lea Environmental Learning Center, 37
 WaterFurnace Envision, 131
Green Cleaning Plan, Sweetwater Sound, 122–125
Green power, energy systems, 85

Habitat for Humanity
 collaborative project, 126–146
 construction techniques/materials, 130–134
 daylight views, indoor quality, 130–132
 energy systems, 128–129, 131
 indoor quality, 130–134
 innovation and design, 127–134
 irrigation, 129
 materials, 128–134
 methods, 127
 NEIGBC, 126, 127–128
 passive cooling, 131
 plan views, 136–144
 stormwater design, 129
 sustainable construction, 126–146
 template, sustainable construction, 145–146
 water efficient landscaping, 129
 windows, 130–132
Habitat, protecting/restoring. *See also* Indoor quality
 LEED® certification, 54–55
Heating, ventilating, and air conditioning (HVAC) systems. *See also* Windows
 energy prerequisites, 76–86
 interconnectedness, 28
 Merry Lea Environmental Learning Center, 35–37
 optimizing energy performance, 80–82
 sustainable construction template, 146
 Sweetwater Sound, 76–86
Heat islands, LEED® certification, 58, 113
Hobbes, Thomas, 9–10
HVAC systems. *See* Heating, ventilating, and air conditioning systems

Ice storage, innovation and design, 112–113
Imbalance, interconnectedness, 5–6
Indoor quality, 97–111
 adhesives, 100
 agrifiber products, 102
 air, 97–103
 ASHRAE standards, 97–99
 carpet systems, 101
 chemicals/pollutants, 102–103
 coatings/paints, 100–101
 composite wood, 102
 construction Indoor Air Quality (IAQ) Management Plan, 99–100
 construction techniques/materials, 130–134
 daylight views, 104–106, 130–132
 ETS, 97–98
 Habitat for Humanity, 130–134
 Indoor Air Quality (IAQ) Management Plan, 99–100
 LEED® certification, 97–111
 light controllability, 103
 low-emitting materials, 100–102

materials, 100–102, 130–134
outdoor air delivery, 98–99
paints/coatings, 100–101
pollutants/chemicals, 102–103
prerequisites, 97–98
sealants, 100
smoke, tobacco, 97–98
Sweetwater Sound, 97–111
thermal comfort, 104, 108–111
ventilation, 98–99
views, 104–106
Innovation and design, 112–125
  Accredited Professional, 115
  chillers, 112–113
  electricity usage, 112–113
  Habitat for Humanity, 127–134
  ice storage, 112–113
  LEED® certification, 112–125
  LEED Accredited Professional, 115
  water use reduction, 73–74, 113
Innovative wastewater technologies, 72–74
  LEED® certification, 72–74
Inspiring spaces, 29
  whole systems thinking, 25–30
Integrated design charrettes, 28–29, 31–38, 128–129
  whole systems thinking, 33
Intellect, 10–11
Interconnectedness, 3–7. *See also* Unconnectedness
  architecture, 26–29
  communication, 26–29
  consequences, 4–7
  HVAC systems, 28
  imbalance, 5–6
  reconnecting to nature, 11–12
Interdependency of organisms
  biomes, 43–46
  ecosystems, 43–46
  nitrogen cycle, 45
  nutrient chain, 45
  sites, 43–46
  water cycle, 45
Irrigation
  Habitat for Humanity, 129
  water resources, 71–72, 129

Landscaping, sustainable. *See* Sustainable
  landscaping
Laws of thermodynamics, 5–6
Leadership in Energy and Environmental Design
  (LEED®) certification
  Accredited Professional, 115
  alternative transportation, 52–54
  bicycle storage, 53
  brownfield redevelopment, 52
  certification process, 48–49
  changing rooms, 53, 54
  communication, 116–121
  community connectivity, 52
  construction activity pollution prevention, 50
  density of development, 52
  development density, 52
  energy systems, fundamental commissioning, 78
  fuel-efficient vehicles, 53

habitat, protecting/restoring, 54–55
heat islands, 58, 113
indoor quality, 97–111
innovation and design, 112–125
innovative wastewater technologies, 72–74
LEED Accredited Professional, 115
levels, 49
light pollution reduction, 59–60
low-emitting vehicles, 53
materials, 87–96
Merry Lea Environmental Learning Center,
  31–32
minimum energy performance, 79
open space, maximizing, 55–56
optimizing energy performance, 80–82
parking capacity, 54
pollution prevention, 50, 59–60
public access, transportation, 52–53
publicity, 116–121
refrigerant management, 79
resources, 87–96
scorecard, 62–66
site development, 54–56
site selection, 51–52
stormwater design, 56–57
sustainable sites, 50–60
Sweetwater Sound, 48–125
transportation, alternative, 52–54
wastewater technologies, innovative, 72–74, 113
water efficiency, 71–74
water efficient landscaping, 71–72
water resources, 71–75
water, stormwater design, 56–57
water use reduction, 73–74, 113
Learning organizations, collaboration, 33–35
LEED® certification. *See* Leadership in Energy and
  Environmental Design (LEED®) certification
LEED Accredited Professional, innovation
  and design, 115
Light controllability, indoor quality, 103
Light, daylight views, indoor quality, 104–106,
  130–132
Light pollution reduction, LEED® certification,
  59–60
Living systems, ecology, 6–7
Lloyd's Crossing
  cost, 19
  'energy means', 19–20
  generative design, 16–21
  'water means', 17–18
  water resources, 17–19
Low-emitting materials
  adhesives, 100
  agrifiber products, 102
  carpet systems, 101
  coatings/paints, 100–101
  composite wood, 102
  indoor quality, 100–102
  paints/coatings, 100–101
  sealants, 100
Low-emitting vehicles
  LEED® certification, 53
  transportation, 53

Materials, 87–96. *See also* Resources
  certified wood, 94–95
  composite wood, 102
  embodied energy, 87–88
  Habitat for Humanity, 128–134
  indoor quality, 100–102, 130–134
  LEED® certification, 87–96
  low-emitting materials, 100–102
  recyclables, 88, 90–92
  regional materials, 93–94
  renewable, 94
  rubberwood, 133–134
  waste management, 90–91
  wood, 94–95, 102, 133–134
Measurement & verification, energy systems, 85
Mental models, collaboration, 34–35
Merry Lea Environmental Learning Center, 31–39
  geothermal unit, 37
  HVAC systems, 35–37
  LEED® certification, 31–32
  sustainable features, 35–38
Mind functions, 11
Minimum energy performance, LEED® certification, 79
Muir, John, 15–16

Nature's conscious representatives, 9–14
NEIGBC. *See* Northeast Indiana Green Build
    Coalition
Neighbor interviews, site research, 47
Nitrogen cycle
  biomes, 45
  ecosystems, 45
  environmental issues, 45
  interdependency of organisms, 45
  sites, 45
Northeast Indiana Green Build Coalition (NEIGBC),
    126, 127–128
Nutrient chain
  biomes, 45
  ecosystems, 45
  environmental issues, 45
  interdependency of organisms, 45
  sites, 45

Occupational Safety and Health Administration
    (OSHA), lessons from, 28–29
Open space, maximizing, LEED® certification, 55–56
Optimizing energy performance
  ASHRAE Advanced Energy Design Guide, 80–82
  energy systems, 80–82
  eQUEST software, 80–82
  HVAC systems, 80–82
  LEED® certification, 80–82
  software, 80–82
Orienting buildings, 76
  sustainable construction template, 145–146
OSHA. *See* Occupational Safety and Health
    Administration
Outdoor air delivery, indoor quality, 98–99

Paints/coatings
  indoor quality, 100–101
  low-emitting materials, 100–101
Parking capacity, LEED® certification, 54

Passive cooling
  buildings, 76–78
  ERVs, 77
  Habitat for Humanity, 131
  Sweetwater Sound, 76–77
Personal attributes, collaboration, 33–37
Personal mastery, collaboration, 34
Photo documentation, site research, 47
Photosynthesis, 6
Plan views
  Habitat for Humanity, 136–144
  Sweetwater Sound, 67–69
Pollutants/chemicals, indoor quality, 102–103
Pollution prevention
  construction activity, 50
  LEED® certification, 50, 59–60
  light pollution reduction, 59–60
Professional, Accredited
  innovation and design, 115
  LEED® certification, 115
Project organization, conventional, 25–29
Public access, transportation, LEED® certification, 52–53
Publicity. *See also* Communication
  LEED® certification, 116–121
  Sweetwater Sound, 116–121

Reconnecting to nature, 11–12
Recyclables
  collection, 88
  materials, 88, 90–92
  resources, 88, 90–92
  storage, 88
Reflective conversation, collaboration, 35
Refrigerant management
  energy systems, 79, 83–84
  LEED® certification, 79
Regional materials, 93–94
Renewable energy. *See also* Energy systems
  on-site renewable energy, 82
  solar energy, 48
  Sweetwater Sound, 76–86
  wind energy, 48
Renewable materials, 94
Research, site. *See* Site research
Residential project. *See* Habitat for Humanity
Resources, 87–96. *See also* Materials
  building reuse, 89–90
  certified wood, 94–95
  LEED® certification, 87–96
  recyclables, 88, 90–92
  waste management, 90–91
Reusable house, Dymaxion House, 22–23
  rubberwood, 133–134
Rubberwood, cabinetry material, 133–134

'scientific method', 10, 13
Scorecard, LEED® certification, 62–66
Sealants
  indoor quality, 100
  low-emitting materials, 100
Site development
  habitat, protecting/restoring, 54–55
  LEED® certification, 54–56
  open space, maximizing, 55–56

Site plan, Sweetwater Sound, 67–69
On-site renewable energy, energy systems, 82
Site research, 46–48
 area surrounding site, 47
 data collection, 46–48
 energy mapping/modeling, 47–48
 floodplain search, 46–47
 neighbor interviews, 47
 photo documentation, 47
 site visits, 47
 software, modeling, 48
 solar energy, 48
 topographical surveys, 47
 trees, 47
 wetland search, 46
 wind energy, 48
Sites, 43–69
 biomes, 43–46
 ecosystems, 43–46
 environmental issues, 43–46
 interdependency of organisms, 43–46
 nitrogen cycle, 45
 nutrient chain, 45
 research, 46–48
 water cycle, 45
Site selection, LEED® certification, 51–52
Site, sustainable construction template, 145
Site visits, site research, 47
Smoke, tobacco, indoor quality, 97–98
Software
 eQUEST software, 80–82
 optimizing energy performance, 80–82
Software, modeling, site research, 48
Solar energy, site research, 48
Storage, recyclables, 88
Stormwater design
 Habitat for Humanity, 129
 LEED® certification, 56–57
 quality control, 57
 quantity control, 56
Survey, thermal comfort, 108–111
Sustainability in action, collaboration, 31–39
Sustainable construction
 approaches, 145–146
 template, 145–146
Sustainable construction, collaborative project.
  See Habitat for Humanity
Sustainable features, Merry Lea Environmental
  Learning Center, 35–38
Sustainable landscaping, 70–75
 water efficient landscaping, 71–72
 water resources, 70–75
Sustainable sites, LEED® certification, 50–60
Sweetwater Sound, 48–69. *See also* Leadership in
  Energy and Environmental Design (LEED®)
  certification
 communication, LEED® certification, 116–121
 cooling, passive, 76–77
 energy systems, 76–86
 Green Cleaning Plan, 122–125
 HVAC systems, 76–86
 indoor quality, 97–111
 LEED® certification, 48–125
 passive cooling, 76–77

 plan views, 67–69
 publicity, LEED® certification, 116–121
 renewable energy, 76–86
 scorecard, 62–66
 site plan, 67–69
 windows, 97–98, 105

Template, sustainable construction, 145–146
Thermal comfort
 indoor quality, 104, 108–111
 survey, 108–111
Tobacco smoke, indoor quality, 97–98
Topographical surveys, site research, 47
Transaction, 10–11
Transportation
 alternative, 52–54
 bicycle storage, 53
 changing rooms, 53, 54
 fuel-efficient vehicles, 53
 LEED® certification, 52–54
 low-emitting vehicles, 53
 parking capacity, 54
 public access, 52–53
Trash, 21–22
Trees, site research, 47

Unconnectedness, 10–13. *See also* Interconnectedness
 conventional project organization, 25–29

Ventilation. *See also* Heating, ventilating, and air
  conditioning (HVAC) systems; windows
 ERVs, 77
 indoor quality, 98–99
 outdoor air delivery, 98–99
Views, indoor quality, 104–106

Waste management
 construction waste management, 90–91
 materials, 90–91
 resources, 90–91
Wastewater technologies, innovative, 72–74, 113
 LEED® certification, 72–74, 113
Water cycle
 biomes, 45
 ecosystems, 45
 environmental issues, 45
 interdependency of organisms, 45
 sites, 45
Water efficiency, LEED® certification, 71–74
Water efficient landscaping
 Habitat for Humanity, 129, 130
 LEED® certification, 71–72
WaterFurnace Envision, geothermal unit, 131
'water means'
 living within, 17–18
 Lloyd's Crossing, 17–18
Water resources, 70–75
 effluent, 71
 irrigation, 71–72, 129
 LEED® certification, 56–57, 71–75
 Lloyd's Crossing, 17–19
 shortages, 70
 stormwater design, 56–57, 129
 sustainable construction template, 146

Water resources (*continued*)
  sustainable landscaping, 70–75
  wastewater technologies, innovative,
    72–74, 113
  water efficient landscaping, 71–72, 129, 130
  water use reduction, 73–74
Water use reduction, 73–74
  innovation and design, 113
  LEED® certification, 73–74, 113
Wetland search, site research, 46
Whole systems thinking, 25–30
  architecture, 25–30
  communication, 25–30

  inspiring spaces, 25–30
  integrated design charrettes, 33
Wind energy, site research, 48
Windows. *See also* Heating, ventilating, and air
    conditioning (HVAC) systems
  ASHRAE standards, 97
  automated, 77
  Habitat for Humanity, 130–132
  Merry Lea Environmental Learning Center, 37
  Sweetwater Sound, 97–98, 105
Wood
  certified wood, 94–95
  composite wood, 102